BARHOLM CASTLE

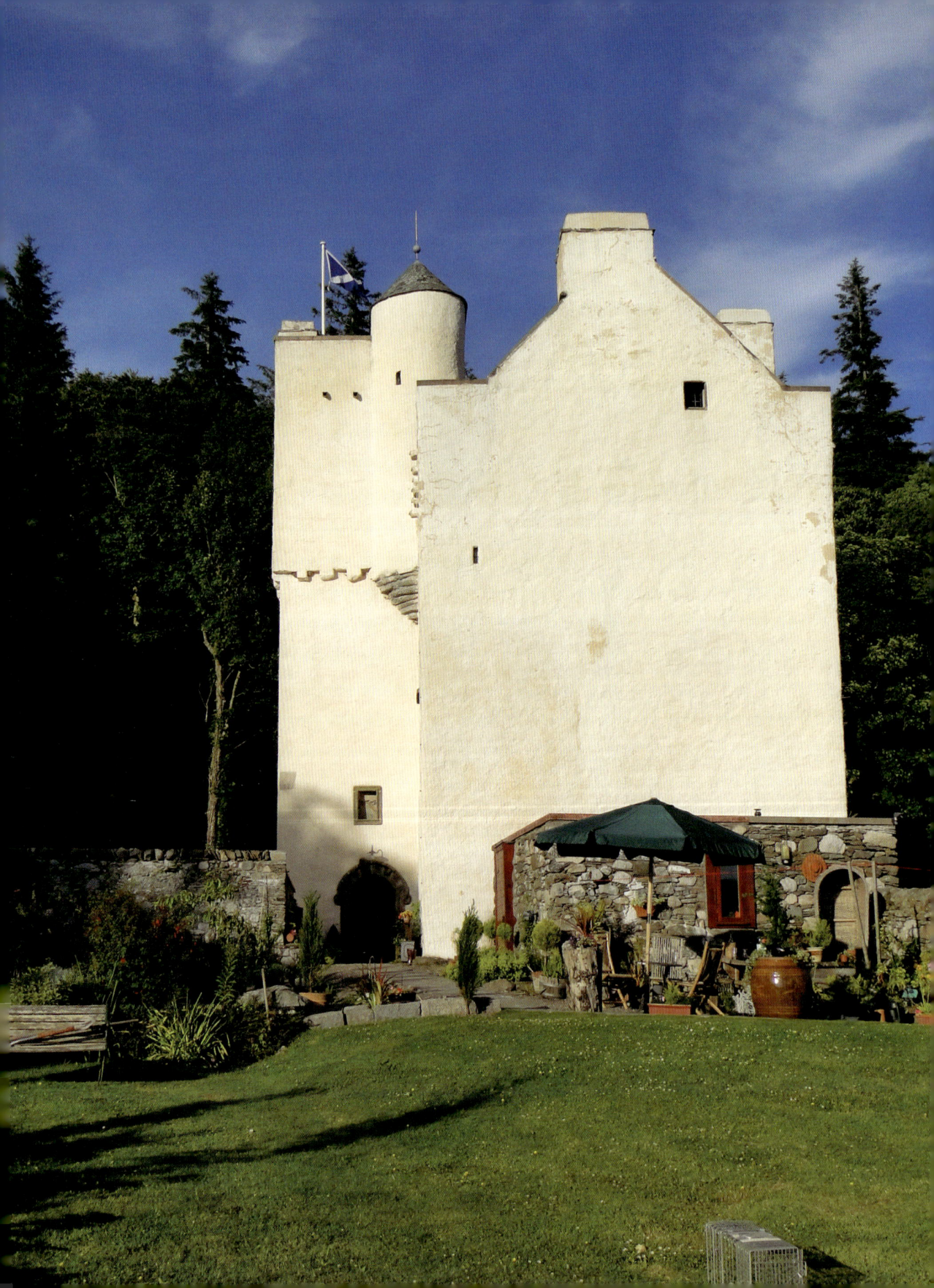

BARHOLM CASTLE

The History of a Home and the Making of a Garden

INCLUDING A HISTORY OF THE BARHOLM McCULLOCH FAMILY

Janet Brennan-Inglis

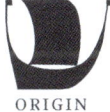

ORIGIN

First published in Great Britain in 2025 by
Origin, an imprint of Birlinn Ltd

West Newington House
10 Newington Road
Edinburgh
EH9 1QS

www.birlinn.co.uk

ISBN: 978 1 83983 087 7

Copyright © Janet Brennan-Inglis 2025

The right of Janet Brennan-Inglis to be identified as the author of this work has been asserted by her in accordance with the Copyright, Designs and Patents Act, 1988

All rights reserved. No part of this publication may be reproduced, stored, or transmitted in any form, or by any means, electronic, mechanical or photocopying, recording or otherwise, without the express written permission of the publisher.

British Library Cataloguing-in-Publication Data
A catalogue record for this book is available on request from the British Library

Designed and typeset by Mark Blackadder

Printed and bound in Britain by
Bell & Bain Ltd, Glasgow

Contents

Preface	vii
Acknowledgements	ix
Picture Credits	x
Abbreviations	xi

Introduction — 1
    Life in Rural Scotland in the Sixteenth Century — 1
    Life within the Castle — 6
    The Galloway Castles — 10
    Life in Galloway in the Seventeenth and Eighteenth Centuries — 18

PART 1   The History of Barholm Castle — 23
    The Building — 24
    The McCullochs of Barholm — 36
    McCulloch and Barholm Characters — 52
    Barholm Castle in the Nineteenth and Twentieth Centuries — 66

PART 2   The Restoration of Barholm Castle — 73
    The First Four Years: Before the Work Could Start — 75
    The Costs and the Building Process — 78
    The Interior — 84
    The Exterior — 99
    A New Life for Barholm Castle — 102

PART 3   The Development of Barholm Castle Garden — 111
    Introduction — 112
    Designing and Creating the Garden — 113
    The Areas of the Garden — 124
    Managing the Garden — 134
    Garden Structure and Features — 151
    Reflections on Gardening — 169
    The Seasons in the Garden — 173
    Changing Times in the Garden — 179

Glossary	185
References	187
Further Reading	189
Index	191

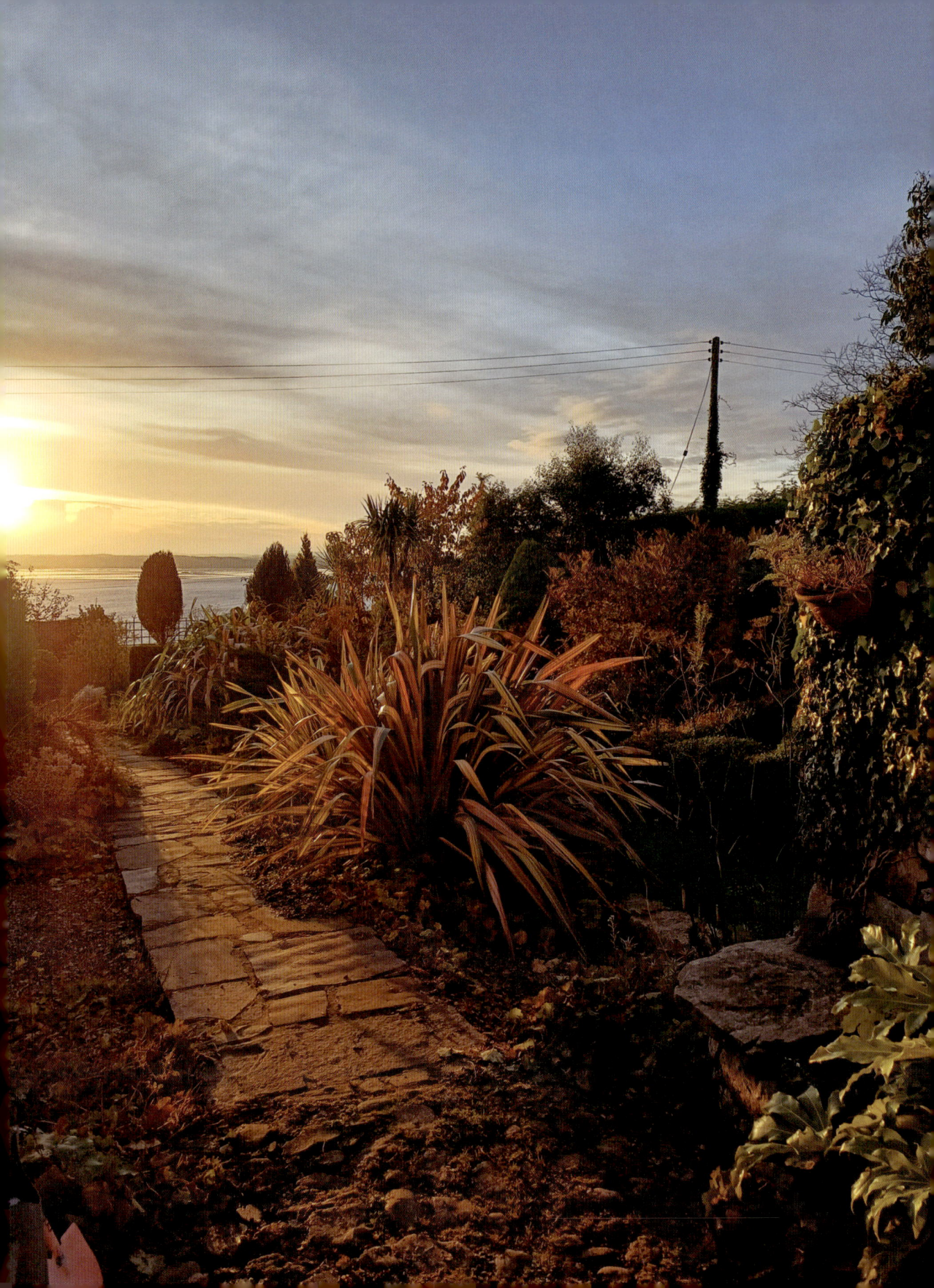

Preface

Ever since we moved to Barholm Castle in 2011, I have intended to write a full account of the history of the building and its owners. Thanks to previous researchers, various visitors from the USA and Canada and the former owner, whose papers included his very helpful historical research, I have been able to put together a history of Barholm Castle and the McCulloch family who built it that I believe is as exhaustive as it currently can be. It has been fun, fascinating and at times frustrating, trying to make sense of the sometimes contradictory and often mysterious pieces of evidence.

The book is in three main parts. In the Introduction I have set the scene in terms of what life was like in sixteenth-century Galloway and, more specifically, within the walls of a tower house. I have also looked at the context of the forty-two other tower houses and castles dotted across Galloway, mostly built at around the same time as Barholm Castle.

Part 1 tells the history of the building of Barholm Castle and of the McCullochs who built it and occupied it for a few hundred years. Their story continues once they left Barholm Castle in the eighteenth century, precipitating its slide into ruination; they built themselves a fine Adam mansion in nearby Creetown. Other McCullochs left the south-west of Scotland to make their fortunes in England and overseas. A few of their stories are told here.

In Part 2, I tell the story of the restoration of Barholm Castle, from the ruin that we first saw in 1997 to the completion of the building works in 2005 and beyond, through the years when it was let out to holidaymakers who wanted to 'live like a laird' in a Scottish tower house, to our finally taking up residence in 2011. The before and after photographs show the transformation that took place over the years.

Part 3 tells the story of the development of the garden we have made. I have always loved reading the stories of garden-making told by heroes such as Christopher Lloyd, Roy Strong, Beth Chatto and Tim Smit. Our garden at Barholm is not a great garden in the mould of Great Dixter, The Laskett or Heligan, but it is much loved by visitors. The 3 acres have developed and changed out of all recognition since we started the garden in 2005. The photographs tell the story pictorially, and I have added a narrative that charts the changes and includes some reflections on various gardening-related topics.

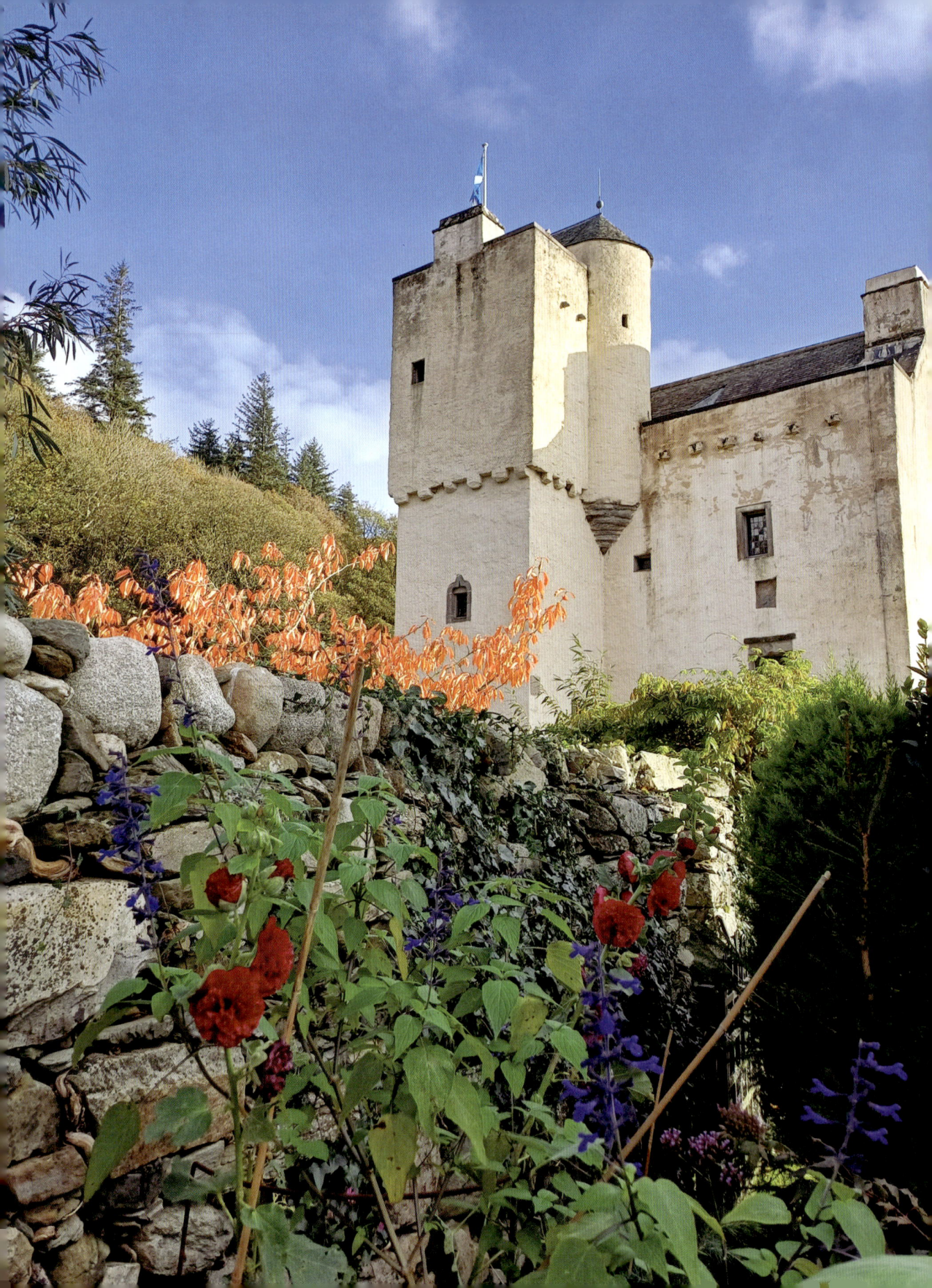

Acknowledgements

Grateful thanks are due to the following people for their help and encouragement:

Dr John Brennan; Dr Rose Brennan; Richard Agnew; Steve and Judy Robinson, executors of former Barholm owner Patrick Whitford; Dr David Hannay of Kirkdale; Dr A. Henry Spong, direct descendant of the McCullochs of Barholm; Adam McCulloch; Mary Winston Nicklin; Everett Post; Peter Drummond, architect; Patrick Lorimer, architect; Michael MacLeod, author and historian; James Shirreff, descendant of Bonnie Bess; Ian Robertson; Doug McCullough, author of *A History of Clan McCulloch* (2025).

Picture Credits

The author and publisher are very grateful to the following individuals and organisations for their kind permission to reproduce the images listed below.

Air Images: aerial photograph of Barholm Castle, 1999, p. 74; aerial photograph of Barholm Castle, 2006, p. 80

Angus Blackburn: the great hall, p. 92; the master bedroom, p. 94; the guest room, p. 96

Andrew Briggs: sketch of Barholm Castle as a ruin, p. 103

Andrew Crawley: the spiral staircase, p. 87; the Brennans at Barholm, p. 108

Heather Davies: watercolour of Barholm Castle, p. 107

Peter Drummond: aerial view of Barholm Castle and garden, p. 34

Clare Hewitt: map of the garden at Barholm Castle, p. 105

Historic Environment Scotland: plans and section of Barholm Castle, p. 33; plan of Barholm House, p. 49; Ian Lindsay plan of Barholm Castle, p. 77

Andy McKean: watercolour of Barholm Castle, p. 106

Michael MacLeod: Miss Jane McCulloch Grant, p. 64; Frederick Wickham Weekes, p. 65; Frederick Wickham Weekes, p. 66; Freddie and Agnes in a motor car, p. 67; intrepid climbers atop Barholm Castle, p. 68; Barholm Castle in the 1930s, p. 69

National Library of Scotland: Timothy Pont's map of Galloway, p. 25; map of Creetown, p. 48; Ainslie map, p. 67

Dr A. Henry Spong: interior of Barholm House, p. 50; Bonnie Bess, p. 56; portraits of Captain Grant and Isabella McCulloch, p. 62

Abbreviations

HES — Historic Environment Scotland, an executive non-departmental public body responsible for investigating, caring for and promoting Scotland's historic environment. HES was formed in 2015 from the merger of government agency Historic Scotland with the Royal Commission on the Ancient and Historical Monuments of Scotland.

HS — Historic Scotland, the predecessor (along with RCAHMS) of HES.

RCAHMS — The Royal Commission on the Ancient and Historic Monuments of Scotland; it was merged with HS in 2015.

RHS — Royal Horticultural Society.

TDGNHAS — The *Transactions of the Dumfries and Galloway Natural History and Antiquarian Society*. These are available online. The first volume was published in 1862.

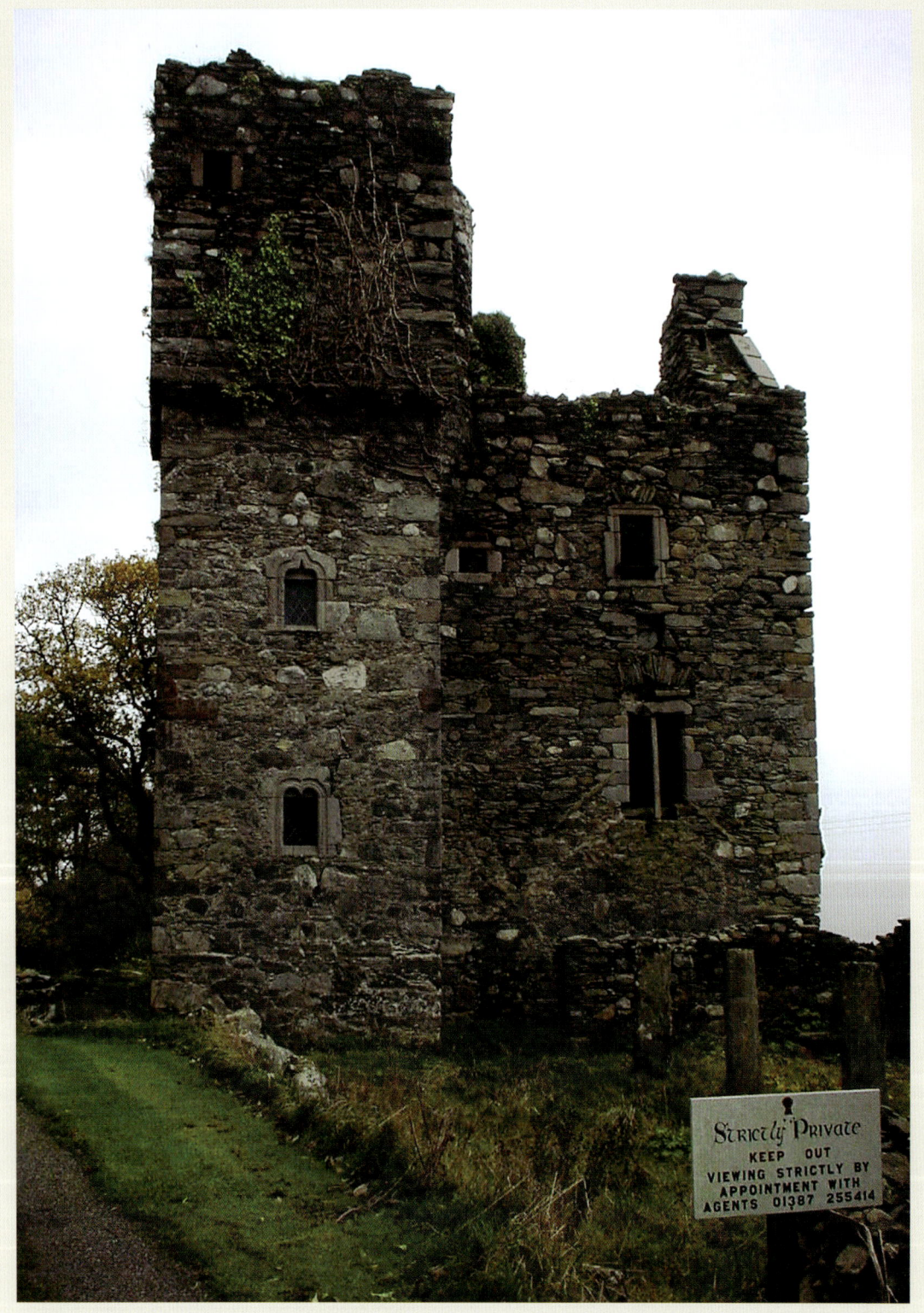

Introduction

When we first saw the ruin of Barholm Castle for sale in 1997 we instantly fell in love with it. We rather naively thought that it would be a fun project to bring it back into use as a domestic residence. After all, it was very cheap at only £65,000. Had we known that it would take two years to buy, four years to get all of the necessary permissions and the design and building teams in place, almost three years to do all the work of restoration, and that we would have to rent it out to holidaymakers for a further five years while we lived overseas, earning enough money to pay for it – the costs trebled from the original prediction, of course – we might have hesitated more. But fools rush in where angels fear to tread. To cut a very long story very short, we bought the place in 1999, and in 2005 the rebuilding of Barholm Castle was finally finished. Only the gardens remained to be developed, which has been our ongoing task for more than twenty years since then – never finished, of course. A garden is a dynamic performance which changes constantly and never, ever reaches a conclusion.

Barholm Castle is most likely a sixteenth-century building. I often think of the inhabitants of five hundred years ago as I move from room to room, up and down the steep spiral staircase, and I wonder about the lives that they led in the tower. The choir that I sing in has a repertoire of sixteenth-century madrigals, which might seem appropriate for the owner of a late mediaeval castle. However, it seems unlikely that the McCullochs of Barholm were sitting around singing madrigals in the tower. They were more likely to have been earnestly discussing their Covenanting beliefs, living an accordingly puritan lifestyle, and perhaps plotting revenge on the neighbours who had murdered members of their family on account of religious differences.

Life in Rural Scotland in the Sixteenth Century

For the majority of inhabitants of sixteenth-century rural Scotland, life was tough. Nearly everyone was involved in agriculture or food production and most people spent their daily existence tending livestock and crops and producing food and drink. They were dependent on the weather, the harvest

Opposite.
Our very first view of Barholm Castle in 1997.

and the fairness or generosity of the laird whose land they farmed. There was a famine in Scotland in the 1590s, and almost half of the years in the second half of the sixteenth century saw local or national scarcity. Difficulties were exacerbated by outbreaks of plague, with major epidemics in the periods 1584–8, 1595 and 1597–1609.

Malcolm McLachlan Harper, in his 1876 book *Rambles in Galloway*, described rural life in the seventeenth century:

> The greater part of the fertile districts of lower Galloway was apportioned to small squatters or crofters, who had neither the means, the inclination, nor the skill to improve the land. They held a right of pasturage in common, on the whole property of the landlord, and the small crofts around their wretched dwellings being the perpetual scene of their agricultural labours, how to improve their material or social condition was never dreamt of by them. From time immemorial this had been the usage, but shortly after the beginning of the last century [i.e. the eighteenth century] various agricultural improvements were commenced.
> (Harper 1876, p. 2)

An excellent representation of sixteenth-century farming life was shown in *Tudor Monastery Farm*, a BBC television series broadcast in 2013, where a historian and two archaeologists lived immersively on a late mediaeval style farm for several months. There had been a previous series where the same team had lived on a Victorian farm for a full year; I was struck by the differences between the two periods in history. The Victorian farm seemed comfortingly familiar – many of the farm tools, household implements and working practices have still been in use within living memory – and life there had a jolly Dickensian quality, despite the lack of modern conveniences. The Tudor farm, on the other hand, was unrelenting hard work with little by way of comfort for the inhabitants; the industrial manufacture of goods had not yet begun, meaning that almost everything had to be hand made from locally sourced materials. Both the agricultural and industrial revolutions of the eighteenth century brought enormous changes and many benefits (as well as misery and disruption) to the people of Scotland, but neither had started when Barholm Castle was built.

Most sixteenth-century dwellings were small and rudimentary, damp and deeply uncomfortable. The tower house, on the other hand, was – literally – the height of relative luxury, although according to Margaret Sanderson,

INTRODUCTION

> The contrast in the standard of the domestic circumstances of landlord and tenant was probably less marked than it was later to become; a tower house might be bigger and safer than the farmer's steading, but it afforded little more *private* accommodation for the householder and his family, in proportion to the number of people living in it. The paternalist feeling, which might extend to physical protection of the tenants by the landlord, was still strong, even in the Lowlands. A laird's children might be fostered in a cottar house, and as the sixteenth century progressed the sons of lairds, tenants and craftsmen would learn side by side in the parish school. (Sanderson 1982, p. 170)

Even so, the lifestyles and possessions of the inhabitants of towers and castles would be much more luxurious than those of the tenants.

In Galloway, the focus of outward communication was towards the Solway estuary. *A Land Apart* is the apt subtitle of Andrew McCulloch's book on the region: 'Even by the seventeenth century the province was relatively isolated from the rest of the country, still an area of warring clans, religious strife, gipsies, smugglers and coastal raiders' (A. McCulloch 2000, p. 12). Roads were generally poor, and the landscape was isolated from the rest of Scotland, penetrable only by high valley passes to the north and east. The sea, however, gave access to Ireland, the Isle of Man, Cumbria and the coast of Ayrshire, as well as providing a means of travelling between towns and villages in Galloway. Just along the coast from Barholm Castle was the Ferry Toun of Cree, officially renamed Creetown in 1792, when John McCulloch V of Barholm had it constituted as a Burgh of Barony. The Ferry Toun was where one could depart – by ferry, of course – for Wigtown, or for the sacred shrine of St Ninian at Whithorn.

In 1525, in an attempt to control the disputes in the lands around the border with England, James V laid down a new Act of Parliament. It required that every landed man in the region 'shall build a sufficient barmkin upon his heritage and lands in the most suitable place, of stone and lime, containing three score foot of the square, one ell thick and six ells high, for the protection and defence of him, his tenants and their goods in troublesome times'. This translates to barmkin walls that were just under 1 metre thick, 5 metres high and enclosing a courtyard of at least 18 square metres. The area would, of course, depend on status of the family constructing the building and the depth of their pockets. Some of Barholm Castle's enclosure walls are almost 1 metre thick at the base, but nowhere near 5 metres high. The one surviving barmkin in Galloway is at Hills Tower near Dumfries, where the walls are about 2.5 metres high; this is about the standard height for the few remaining barmkins in Scotland, such as those at Smailholm and Craigievar castles. There is clear evidence at Lennox Plunton Tower near Gatehouse of Fleet that its barmkin was 2.74 metres high. Only military buildings tend to have walls higher than 5 metres.

Larger establishments than Barholm would have had extensive outbuildings housing stables, an alehouse, a dairy and a bakehouse, but as a small laird's tower, Barholm would have needed only a few agricultural outbuildings, including a small dairy and bakehouse. The inhabitants of the castle would have been supported by a range of rural craftsmen, including a blacksmith, wheelwright, cooper, weaver, shoemaker, miller, baker and tailor. To sustain themselves in such a rural area, these men would have divided their time between farming and the practice of their craft. They were in thrall to the rhythm of the seasons and the weather, working from sunrise to sunset. In the depths of winter, the sun rises after 8 a.m. and sets before 4 p.m., so the days would be short. But in the west of Scotland, summer days are long and the climate is temperate, giving an opportunity to make up the time lost in the dark winter months. Ploughing with cows or oxen was a task that the majority of men had to engage in. It was very hard physical work and required a degree of skill to complement brute strength.

The relationship between laird and tenant farmers is explored by Margaret Sanderson:

> The contact between tenants and landlords, most of whom, after all, were lairds, bonnet lairds and feuars, was probably closer in the sixteenth century than it was ever to be again, if for no other reason than that landlord and tenant still spoke the same language, vernacular Scots. Absenteeism, at least among the smaller lairds and feuars, was probably the exception, at any rate for long periods. (Sanderson 1982, pp. 170–1)

'Man works from sun to sun, but a woman's work is never done'

Women had many tasks to complete beyond childcare. They had to milk the cows and goats, make the cheese, daily bread and ale, look after the poultry and pigs, clean the house, wash the linen, cook the meals, tend the herb and vegetable garden and generally make sure that the household ran smoothly. Most of these tasks could not be alleviated by any kind of labour-saving device, because very few had yet been invented, but general help would have been given by the older children of the household. One of the most common causes of death for women – apart, of course, from childbirth – was their heavy linen skirts catching fire as they stood cooking over an open fire.

The food

The sixteenth-century diet was dictated by the seasons and by the religious calendar of feasting and fasting. During Lent, in February or March, when no meat, eggs or dairy produce were to be consumed, dried peas and beans were made into savoury puddings, supplemented with the staples of bread and ale. Even for those who did not care to follow the Church's injunctions, after a long winter the hens and geese were often not laying, the cows not producing milk and the beasts scraggy, so there was little to offer by way of animal protein. There might be some pickled herring or salted fish from last year, but fresh fish was unlikely as fishing tended to be a seasonal activity. By Easter, the most important feast of the Christian year, the grass would be growing again, the poultry laying and cows and sheep producing milk. Milk was also used to produce cheese, and in order to start that process, a young calf would need to be slaughtered to provide rennet. The meat from the calf would provide an Easter meal, the first meat of the year for most. In April, stillborn lambs provided bursts of meat-eating, which could be flavoured with fresh herbs. By summer, after the sheep were shorn, mutton would be available, and eggs, butter, milk and cream would continue to supplement the diet, along with fresh strawberries and some green leafy vegetables and fresh peas and beans. In the autumn, the grain harvest meant that pies, pastries and cakes would be plentiful; beef would also be at its best. November was the traditional time for slaughtering pigs, so sausages, hams and bacon would be produced, with some stored for the winter. Herring and cod would be pickled, also to store for the winter, and hard cheeses made which kept for weeks or months. At the beginning of December came Advent and four weeks of abstinence from meat, although fish was allowed. Turnips, parsnips and carrots provided bulk to the pottage made throughout the month. The feast of Christmas might be celebrated with a pickled pig's head and mince pies made with meat, suet and dried fruits. In January the stored cheeses, pickled fish and cured meats, as well as some sprouting scraps of winter kale and dried peas and beans – and, of course, the year-round staples of bread and ale – would provide a reasonably healthy diet until Lent began the cycle again.

Life within the Castle

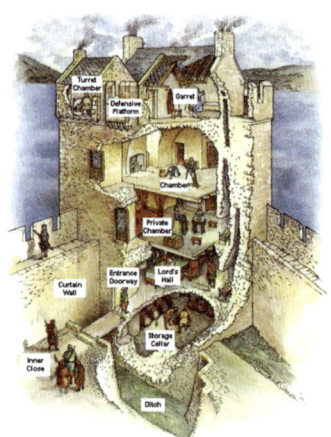

Cutaway section of Urquhart Castle: this shows a typical layout for a Scottish mediaeval tower house, and the thickness of the walls.

Like most late mediaeval buildings, Barholm Castle is a kind of Tardis in reverse. It looks imposingly grand, and people assume that it must have commodious accommodation. But the walls are so thick that there is less living space inside than you might imagine. It is the very opposite of the estate agent's description 'deceptively spacious': the footprint of the ground floor is two thirds wall and only one third room. The vaulted cellar, which has become our kitchen, and which constitutes the ground floor of most tower houses, would have been a storehouse and not in general use by the family.

Interiors of sixteenth-century tower houses had little natural light, because of their small windows, but they were usually brightly decorated. Walls were either panelled, often with paintings on the panelling, or hung with tapestries, which also added a layer of insulation. Ceilings were often brightly painted. Textiles such as rugs and runners were richly patterned, and ornamental ceramics would add to the impression of colour and pattern. The great hall was where family activity was centred. With its huge fireplace and window seats, there was plenty of space for dining and family activities. There may have been a screen at one end to give the laird and his family some privacy; this was often the layout in great halls. Rushes would have been laid on the floor, to absorb food spillages and mud from animal paws and human feet.

The great hall was also the focus of hospitality, where the laird's family and their guests and servants dined. Michael Pearce has made extensive studies of sixteenth-century Scottish inventories, which shed light on the furnishings and possessions of the lairdly class.

> The sixteenth-century hall was equipped for a ritual of hospitality which was instantly recognisable and socially cohesive. Its hierarchic arrangement was intended to emphasise inequality of wealth and power. Followers were intended to be reassured by the display that their masters would fulfil social obligations towards them. Six characteristic groups of furnishings are regularly seen in sixteenth-century inventories:

- the high table, the principal seat and its decoration;
- the 'side tables', seating for everyone else;
- the cupboard and silver plate;
- the hanging chandelier, often described as a 'hart horn';
- displayed weapon(s) – often a single halberd as a symbol of seigniorial power;
- the cloths and napery used during dining. (Pearce 2015, p. 80)

The lairdly inhabitants of the sixteenth-century tower house often had plenty of personal belongings. The thorough inventory of clothing and household articles from Elshieshields Tower near Dumfries (which is a similar size to Barholm), 'left at the decease of Lady Elshieshields 7th December 1670', shows that the possessions of a local lady in the seventeenth century were extensive. Among her 'clotheses' were twelve pairs of shoes, two pairs of boots and a pair of galoshes. She left nine headdresses, fifteen shirts, seven pairs of stockings, a chapeau and a hankie. Among her household articles were forty-two pewter plates, twenty-nine small linen sheets, thirty-two napkins and twelve new blankets. The range of goods in a sixteenth-century tower house would have been similar, although Lady Elshieshields does sound rather extravagant.

Sleeping arrangements were probably crowded. If there were many family members and servants to accommodate, some people would sleep in the great hall. On the second floor of Barholm Castle were two bedchambers, each with its own door accessed by a little corridor built into the thickness of the wall. Each chamber had a fireplace and two windows. Both of the fireplaces were modified at some point to make them smaller, with stones inserted at the sides and a new, lower lintel, presumably to save on fuel. If the laird and his wife had a bed with curtains, that would have afforded them some additional privacy and protection from drafts.

One of the chambers on the main bedroom floor (now used as a bathroom) has a garderobe toilet with a 'moss box' cut into the wall for the mediaeval equivalent of toilet paper. There is even a little window in the garderobe chamber. At the east end of the exterior south wall, the exit of the garderobe chute can clearly be seen at ground level, with a large flat stone set in at an angle of 45°, no doubt to aid the efficient egress of effluent. Those who had the unpleasant task of collecting human excrement from garderobes were called 'gong farmers', and large castles would regularly employ such workers to manage and empty these cesspits. Barholm was probably too small an establishment to need such services and has no cesspits; a general low-ranking servant would have attended to matters when necessary. The other bedchamber also has a similar tiny apartment, or intra-mural recess with a window and door, which, mysteriously, contains no garderobe. It may have been a 'dry stool', 'close stool' or 'stool of ease' chamber, which contained a chamber pot inside a cabinet or inside a stool. But why one room should have a garderobe and the other not is a mystery. The empty intra-mural chamber may instead have been a muniments room, used to store title deeds and other important documents, perhaps with an iron door to protect the contents from the danger of fire. A second garderobe chute can be seen outside the castle at the south-west corner, but this does not line up with the intra-mural chamber; evidence shows that this originated from the wall walk (a passageway behind the castle wall) outside the long gallery on the floor above.

The third floor, as was common in such towers, was one long gallery, or garret (see page 6), with a fireplace and small window at one end. There is also evidence that there was a garderobe near the window, perhaps on the wall walk, with a step to the outside. This would be where servants slept, with the highest-ranking nearest the fireplace and window. There may have been dormer windows bringing in more light, but there is no evidence one way or another as the roof disappeared more than two hundred years ago.

Living in the castle for several years has given us insight into the likely lifestyles of the former inhabitants. When we had a lengthy power outage one winter, we discovered that without the underfloor heating the great hall becomes cold very quickly once the fire has died down. With the full width of the great hall fireplace reaching 9 feet, the amount of wood needed to keep a fire going would have been staggering. We know that the woods around the castle were mainly of oak and ash, which both burn well and would doubtless have been used in the castle for heating and cooking. We occasionally light a fire at Christmas or when we have guests; a large basket of logs does not last a full evening, and that is in the relatively small grate which we have inserted in the fireplace. Even with the small grate the fire gives out a good heat – as long as you keep close to it.

Tower houses generally had small windows with unglazed wooden shutters. Often these were expanded in later years once glass-making technology had evolved to make larger panes cheaper and more readily available. This did not happen at Barholm, however: instead of being modernised, it was abandoned in the eighteenth century. We had windows with unglazed shutters installed in the great hall at Barholm as there was evidence that this type of window was there originally. The upper part would have been glass or, if that was not available, oiled linen or translucent animal hide. The interior

INTRODUCTION

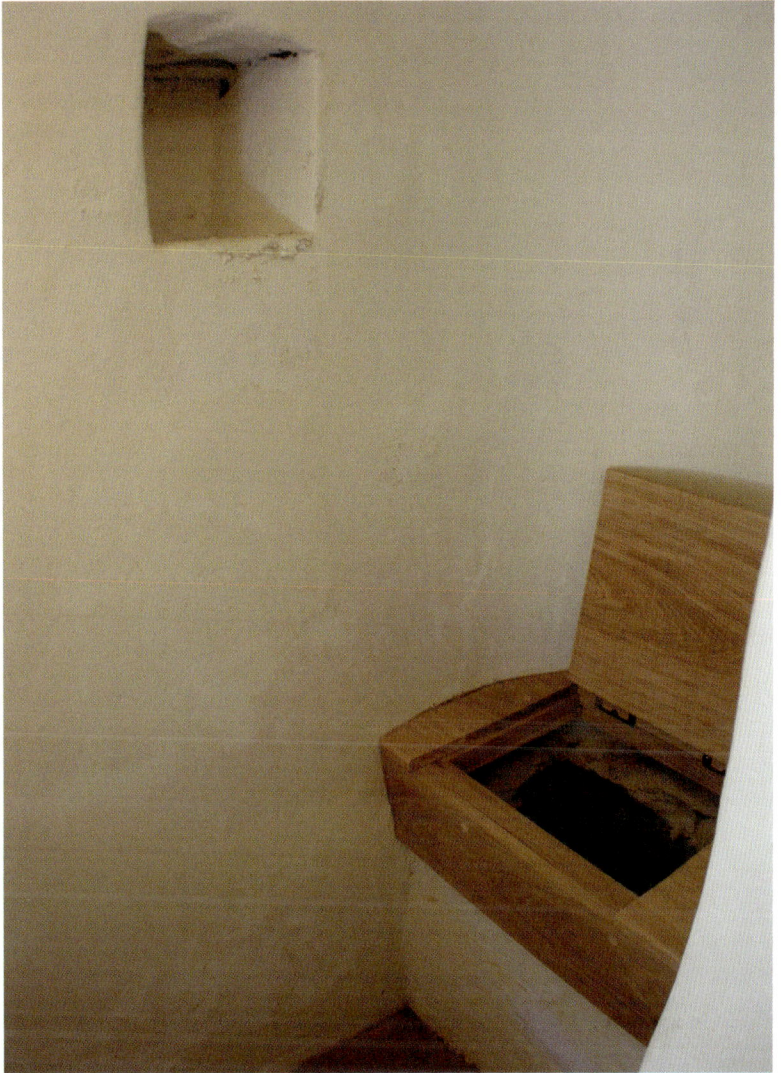

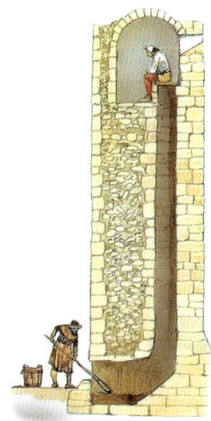

The garderobe on the second floor, with its moss box cut into the wall.

A mediaeval garderobe in action, with the 'gong farmer' shovelling out the fallen excrement.

of the great hall would have been gloomy, even on sunny days. Window seats made the most of natural light, and I imagine the women who were involved in sewing would have used them. After dark, oil lamps or tallow (rendered animal fat) candles would be used to light the castle.

The historian Alastair Maxwell-Irving (2014) notes that there would have been a yett at the door, as there is a bolthole for a drawbar. I keep a selection of secateurs and string in the bolthole, handily placed for snatching up as I go out to the garden.

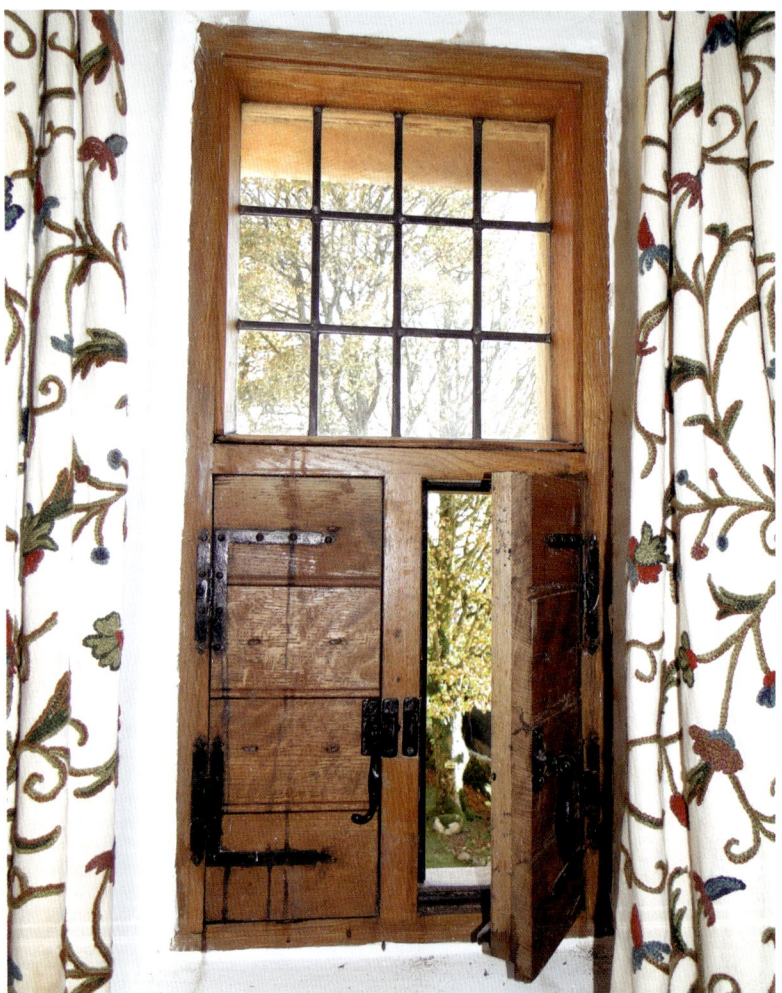

Wooden shuttered window in the great hall, showing the unglazed bottom part and small leaded panes above. These may have been animal hide or waxed linen originally, rather than glass.

The Galloway Castles

All over Scotland, from the middle to the end of the sixteenth century, numerous tower houses were being built, including many of the forty-two in Galloway that are still standing and recorded here. What Stewart Cruden (1981, p. 144) calls the 'long pause' – when few tower houses were built – occurred from about 1480 until the Scottish Reformation in 1560. After the Reformation there was a frenzy of building, leading to an estimated total of 6,000 castles and tower houses across Scotland in the sixteenth century. Castle historians MacGibbon and Ross point out that tower houses were 'remarkably uniform' throughout Scotland: 'in few instances is there any indication of marked originality or individual effort. In other words, there is rarely any appearance of the architectural or designing mind; the work is rather that of the builder acting on traditional lines' (MacGibbon and Ross 1887–92, vol. 5, p. 545). The builder would probably be a master mason, with a group of masons working under him. Indeed, the builder who carried out our restoration work was firmly of the opinion that an architect was still unnecessary in the twenty-first century. However, although there may be little evidence of the 'designing mind', the Galloway tower houses that are more or less contemporaneous with Barholm Castle do display a large range of sizes, settings and levels of verticality, finish and comfort, despite all being recognisably of the same form.

Two local towers were also owned by McCullochs, i.e. Myrton and Cardoness, both now ruinous, and there would have been communion among them. Others, such as Carsluith and Hills, were owned by Catholic families, the Browns and the Maxwells respectively, and they were McCulloch adversaries. Most Galloway towers are ruinous, or at least unoccupied, some depressingly so (e.g. Cally Castle and Castle Clanyard). Only eleven are inhabited (Abbot's, Auchness, Barholm, Barscobe, Rusco, Buittle, Craigcaffie, Hills, Kirkconnell, Lochnaw and Isle of Whithorn) and three (Castle of Park, Machermore and Old Place of Monreith) are let out for holiday accommodation. Many would have been built on the sites of earlier buildings where defensive features were present and there was access to the sea and clean water.

The architectural historian Charles McKean was puzzled by

> those who decided to enhance the *verticality* of their house in the teeth of domestic logic. Why erect a tall house – sometimes an *unconscionably* tall house – when height was not required for defence? Probably because the owners considered height as a way of signalling status and lineage in a society preoccupied with history, genealogy and precedence. Perhaps, too, they were determined that their seat should be visible above the tops of the increasingly protected trees; or even that this was the form of house required by their status or their pocket. (McKean 2001, p. 143)

Certainly, almost all of the Galloway tower houses are uncompromisingly vertical; they make a statement in the landscape and would have been even more prominent five hundred years ago, when they were harled and painted. What has become clear, as I have looked for a common set of dates for the castles of Galloway, is just how little we know of the precise dates of building and the details of ownership and occupancy for many of the individual tower houses. 'Probably' is the most common and useful adverb I can use in their descriptions.

It is clear from the number of towers and castles in Galloway – over forty of which are still in existence – that Barholm Castle was part of a network of grand buildings across the region, ranging from small laird's houses to the fortified magnificence of Threave. The area of Galloway is 1,418 square miles; given that most castles tended to be clustered along the coast or by rivers or lochs, near villages and towns, the buildings must have been very evident. Most would have been harled and limewashed, like Barholm, and thus would have stood out in the countryside, signalling the status of their lairdly owners. The majority were built, or at least expanded, in the fifty years between 1560 and 1610, making it likely that many of them shared builders and masons. Carsluith Castle is near Barholm and very similar in size and design, and Buittle Castle and Sorbie Tower share characteristics, for example. When Barholm was being restored, our builders were also employed by the owners of Lochnaw Castle near Stranraer, who were restoring their building at the same time.

Galloway castles west of Barholm

Barholm Castle lies almost in the middle of Galloway. Heading westwards from Barholm, just a couple of miles along the coast is the ruin of **Carsluith Castle**, probably built originally in the fifteenth century as a square tower, with the stair tower probably added in 1568. It is very similar in size and style to Barholm but was owned by the Roman Catholic Brown family, whereas the McCullochs of Barholm were staunch Protestants. Carsluith is in the care of Historic Environment Scotland (HES). Another mile or so further on is **Cassencarie** or **Castle Cary**, built between 1575 and 1600, standing in a caravan park and in danger of collapse. Nothing is now left of **Muirfad Castle**, near Palnure, which was still in existence in 1800, but only a site by 1849. **Machermore Castle** near Newton Stewart, baronialised and extended in the nineteenth century, originally dates from the late sixteenth century. The **Old Place of Mochrum** was first built in the early sixteenth century and added to at the end of the sixteenth century. Nearby, also in Mochrum Parish, is **Myrton Castle**, a ruined sixteenth-century tower built on a twelfth-century motte. **Castle Stewart** near Newton Stewart is described by Canmore as

INTRODUCTION

'probably 17th century' but 'its plan and general character are suggestive of an earlier period'.

Garlies Castle near Newton Stewart was a great stronghold of the Earls of Galloway. It is difficult to date because of the ruinous state of the remains, but it seems that the original fourteenth-century keep had a tower added in the sixteenth century and that other ancillary buildings, including extensive stables, were added between these dates. Barholm Castle, as we shall see, had a 'belly' that nearly destroyed it. Garlies was not so fortunate:

> From the state of the ruins it is clear the great contributing cause to the destruction of the keep was the weakness of the wall next the courtyard, pierced as it was with so many doorways and a wide passage. As now uncovered, the central part of this wall and the jambs of the door lean considerably outward, and it is evident the haunch of the vaulting arch on this side must have burst out with damaging effect. Indications of this catastrophe probably led to the abandonment of the building.
> (MacGibbon and Ross 1887–92, vol. 5, p. 282)

Cruggleton Castle near Garlieston, a thirteenth-century stronghold, was already abandoned by 1680 and its stones were used for building material elsewhere. Only a picturesque arch remains. It was a huge fortress when it

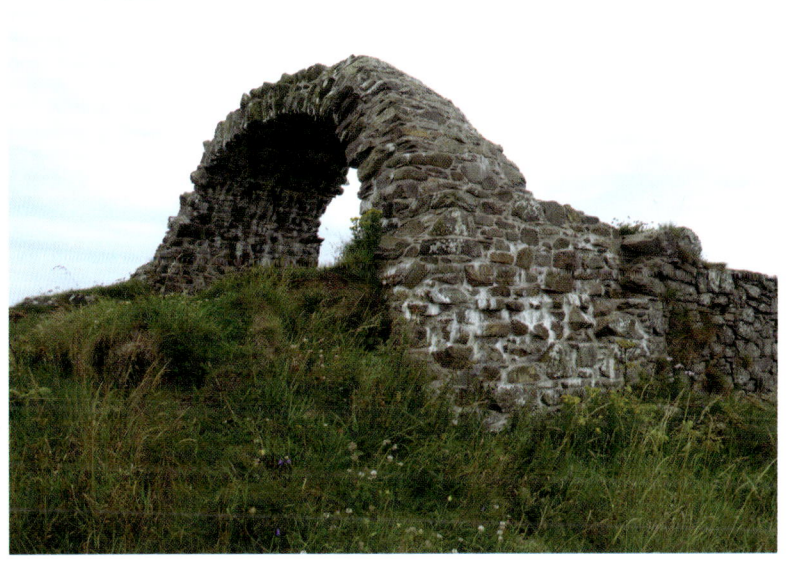

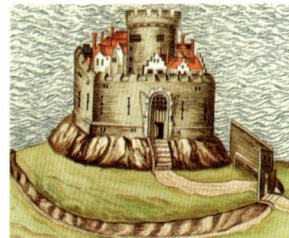

Cruggleton Castle drawn in 1563 when spies of Queen Elizabeth I visited south-west Scotland to examine the defences at several castles.

Left.
All that remains of Cruggleton Castle today.

was in originally built, and, as can be seen in the English 'spy drawing', was a glorious building in the sixteenth century.

Little also remains of **Baldoon Castle** near Wigtown, apart from half of a wall and a pair of ornate Renaissance gateposts, which give an indication of the grandeur of the former building. According to Canmore, it was probably built in the early seventeenth century. **Isle of Whithorn Castle** is claimed by Canmore to have been built in 1674, one of the last tower houses to be built in Scotland. Possibly its remoteness protected it from the vagaries of fashion. In the late eighteenth century, smuggling from the nearby Isle of Man was widespread in the region. The castle was then the residence of the Superintendent of the Coastguard, Sir John Reid.

The **Old Place of Monreith**, also known as **Dowies**, is a small laird's house near Port William, built about 1600. It was restored from dereliction by the Landmark Trust in 1983 and is now available as a holiday let. This has also been the case with **Castle of Park**, also known as **Park House**, near Glenluce Abbey, which was built for Thomas Hay in 1590 and rescued from dilapidation, firstly in the 1970s by the Ministry of Works and again in 1990 by the Landmark Trust. **Castle Kennedy** was built in 1607, to replace an earlier tower. It was destroyed by fire in 1716 and has never been restored. It stands as a romantic ruin in the grounds of fine landscaped gardens that are open to the public. **Sorbie Tower**, although remote, is well-proportioned and was built towards the end of the sixteenth century. It has recently had a temporary roof installed to protect it from further degradation. **Craigcaffie** is another very small laird's tower, near Stranraer, which was built in 1576. **Lochnaw Castle** has been much altered and added to, but its square tower is of the sixteenth century, although undated.

Galdenoch Castle near Stranraer was built sometime between 1547 and 1570. It was still in use as a farmhouse in 1684 but when MacGibbon and Ross visited in 1890 they said, 'It has evidently been abandoned for many years' (1887–92, vol. 3, p. 507). It now stands as a well-preserved ruin in farmland. **Castle Clanyard**, also near Stranraer, has not been so well preserved. Only a part of its west wall remains; the Rev. Andrew Symson, who visited in 1684, described it as 'having been of old a very great house, but now something ruinous'. It was probably built in the late sixteenth century. In Stranraer, the **Castle of St John** dominates the town centre and houses a museum. It is a good example of a sixteenth-century L-plan keep (Canmore); it was originally built in 1500 and extended upwards in the seventeenth century to fit it out as the town jail. **Dunskey Castle**, on a precipitous clifftop site in Portpatrick, originally dates from very early times, but was rebuilt in the sixteenth

INTRODUCTION

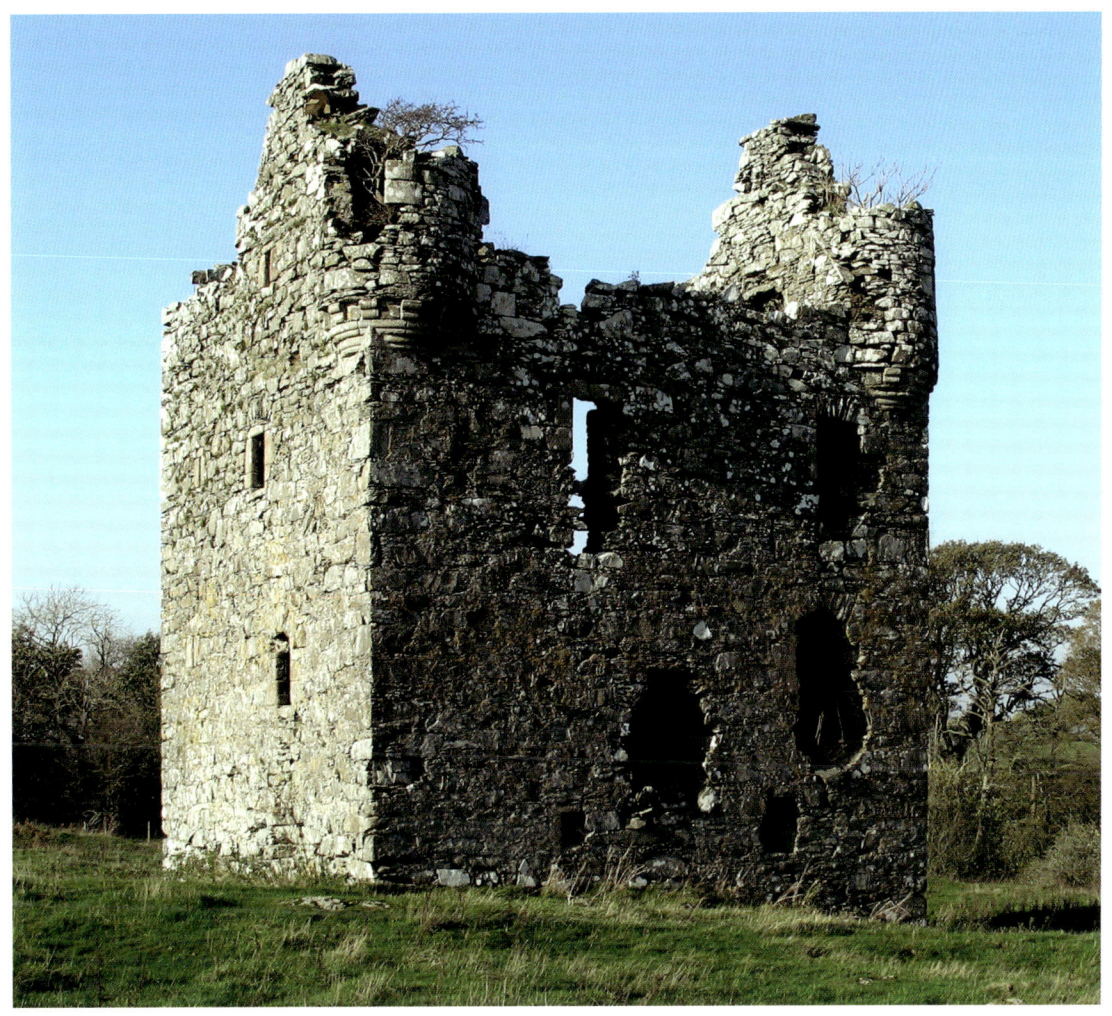

Lennox Plunton Castle – a possible future restoration project?

century. **Balzieland Castle** is now only a tall fragment of a castle, its remaining wall incorporated into the wall of the walled garden of Logan Botanic Garden. It, too, is probably of the sixteenth century. Nearby is **Auchness Castle**, a much-altered sixteenth-century tower, which is still inhabited.

Galloway castles east of Barholm

The castles described above all lie to the west, towards the Irish Sea. To the east, towards Dumfries, are as many again. About five miles along the coast is **Cardoness Castle**, also owned by McCullochs, sometime partners in crime

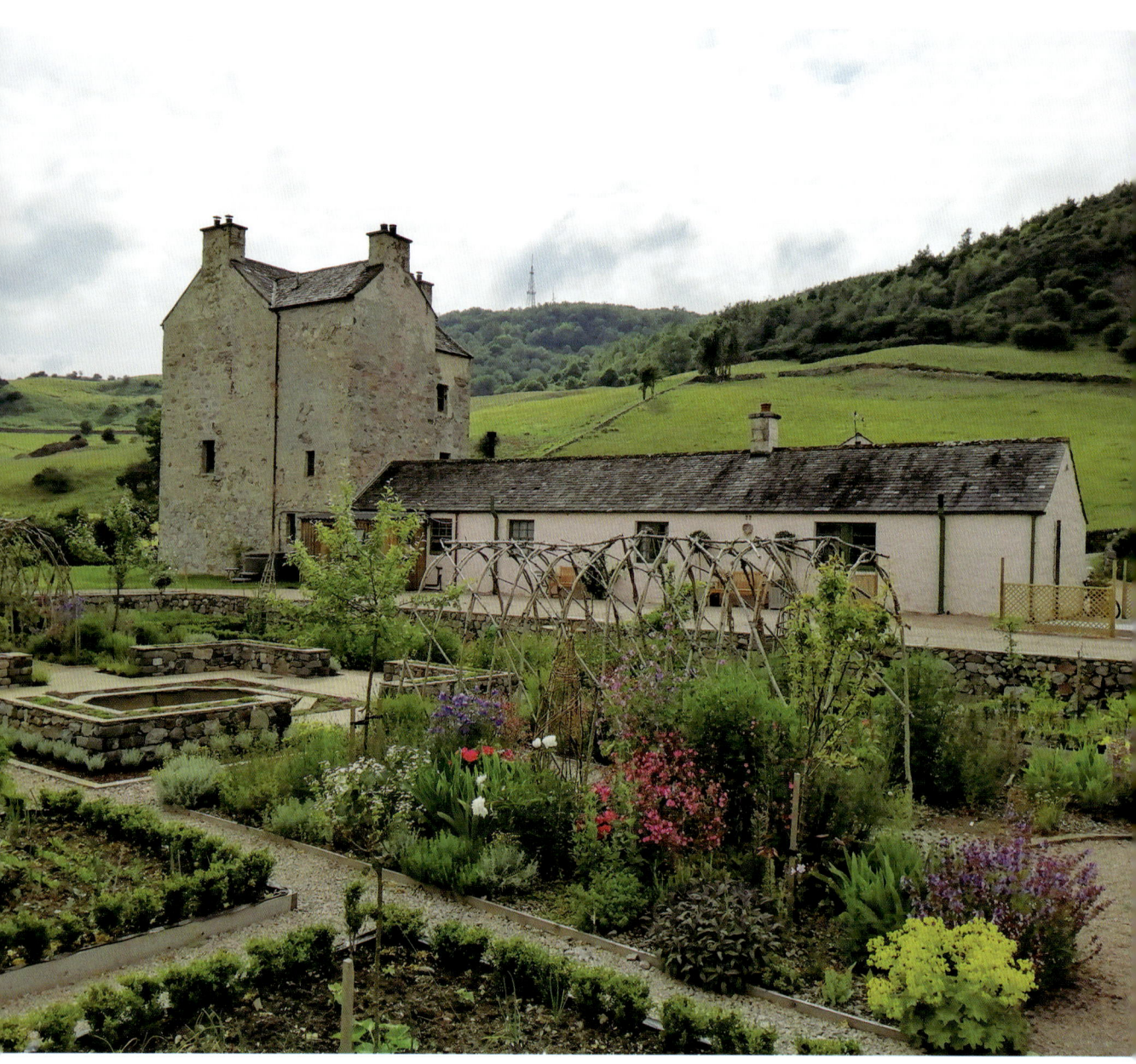

Buittle Castle near Dalbeattie, with its sixteenth-century garden.

of the Barholm McCullochs. It was built in the late fifteenth century and is in the care of HES. **Rusco Tower** to the north of Gatehouse of Fleet was built sometime in the sixteenth century and was restored in the 1970s. **Cally Castle** near Gatehouse of Fleet is too ruinous to be dated, but it would be surprising if it were not also sixteenth century. Also near Gatehouse of Fleet is **Lennox**

Plunton, a tiny tower built around 1575. It is ruinous but was almost restored about twenty years ago and could yet be a restoration project for someone.

Buittle Castle, near Dalbeattie, is a sixteenth-century tower house with a twelfth-century motte-and-bailey site in the grounds. **Auchenskeoch Castle** is another building that dates from the second half of the sixteenth century. It is very ruinous and situated in a farmyard enclosed by modern sheds. **Edingham Castle** was built between 1570 and 1585 and now sits ruinous in farmland.

Maclellan's Castle is a fine nobleman's house dating from the late sixteenth century which stands in the centre of Kirkcudbright; neglected and smothered in ivy up until 1913 it is now cared for by HES and open to the public. Not far from Kirkcudbright, near Twynholm, is **Cumstoun (or Compstone) Castle**, an early sixteenth-century ruined tower that stands in the grounds of Cunstoun House. **Balmangan Tower** is another example of a sixteenth-century ruin, abutted by a farmhouse and only surviving to first-floor level.

Most of the tower houses described so far, despite being called 'castles', have been principally domestic residences with the exception of Garlies, Dunskey and Cruggleton. **Threave Castle,** near the town of Castle Douglas, was primarily a mediaeval defensive structure, situated on an island in the River Dee and built in the fourteenth century by Archibald the Grim, Lord of Galloway. It is in the care of HES, although it belongs to the National Trust for Scotland.

Drumcoltran Tower near Dalbeattie was built sometime after 1550 by Sir John Maxwell and his wife Agnes Herries and has similarities to Buittle. **Orchardton Tower** is Scotland's only round tower house, although this is a common building form in Ireland; it dates from the mid fifteenth century. Little remains of **Barclosh Castle**, which dates from the late sixteenth century. **Kenmure Castle** near New Galloway was an early, probably twelfth-century stronghold and was rebuilt in the late sixteenth century after being destroyed by Regent Moray following the Battle of Langside, then added to several times over the next three centuries. It became ruinous only in the mid twentieth century. **Barscobe Castle** is a very late example of a tower house, built in 1648. MacGibbon and Ross say, 'this is an interesting example of the kind of house which gradually evolved itself out of the old pale tower towards the end of the seventeenth century, and shows the extent of the accommodation which was then considered sufficient for a country laird' (1887–92, vol. 3, pp. 523–4). **Earlstoun Castle** near St John's Town of Dalry dates from the late sixteenth century and was inhabited until the nineteenth century. It was almost restored by the Vivat Trust in the early twentieth century, but the Trust ran out of funds and was dissolved. Earlstoun features in S.R. Crockett's bestselling novel of 1895, *Men of the Moss Hags*. Not far north, very little remains of **Dundeugh Castle** beyond a few feet of broken walling. It, too, is probably of the sixteenth century.

Finally, almost as far east as Galloway stretches, **Abbot's Tower** was built by the Abbot of Sweetheart Abbey in about 1580. It was restored from ruins in the 1990s. Nearby **Kirkconnell Tower** has been much altered and extended over the centuries, but its four-storey square tower dates from the sixteenth century. **Hills Tower** near Dumfries, but still within the Stewartry of Kirkcudbright, was built between 1522 and 1566 and is unusual in that it still has its barmkin and gatehouse intact.

Other Galloway castle restorations

Barholm is not the only castle or tower house to have been restored in Dumfries and Galloway. Lochnaw Castle, as already mentioned, was restored at about the same time as Barholm, although it was not in such a ruinous state. Abbot's Tower near New Abbey, Barscobe Castle, Rusco Tower, Buittle Castle, Craigcaffie Tower, Hills Tower, Kirkconnell Tower, Castle of Park, Machermore Castle, Old Place of Monreith and Old Place of Mochrum have all been rescued from ruin or dereliction. All but the last were restored in the last fifty years. Old Place of Mochrum was restored from a ruinous state and extended by the 3rd and 4th Marquesses of Bute over the period 1873–1912. The 3rd Marquess of Bute also started the restoration of Sanquhar Castle in Dumfriesshire in 1895, but work halted on his death in 1900 and the castle remains perilously ruinous.

Life in Galloway in the Seventeenth and Eighteenth Centuries

Solway smuggling

While Galloway has been depicted as a poor region of Scotland – indeed, it still is – there was one money-making enterprise in the seventeenth and eighteenth centuries that apparently involved all sections of society: smuggling. Although it was illegal, many Scots thought it was a justified way of dodging an unfair tax on goods imposed by a distant English government (after the

Union of the Parliaments in 1707, at any rate). It was also a relatively easy way to make a quick profit and the risks were seen to be worth it, especially when the government's forces were undermanned and often running scared of the smugglers. It seems that every echelon of society in Galloway was involved in the smuggling trade, from peasants to lairds. The minister of Anwoth parish (the neighbouring one to Kirkmabreck, in which both Barholm Castle and Barholm House lie), the Reverend Carson, was dismissed from his post in 1767 for 'smuggling and encouraging that practice'.

Historian Gordon Irving claims that 'The village of Creetown was a centre for smugglers, many of whom made their homes there. Boats used to bring their contraband cargoes into the little port at Carsluith, three miles away' (1971, p. 39). He tells the story of a gang of Creetown smugglers whose haul of salt was seized by the authorities. The smugglers then tricked the soldiers by filling up old salt bags with sawdust and using them as a decoy to lure the soldiers away. While they were occupied in chasing the men with the bags of sawdust, another group of smugglers grabbed the sacks of real salt and hid them in a hole in the garden of Cassencary House, the neighbouring estate to Barholm House. Whether this was done with the knowledge or connivance of the laird is not known, but it is likely that he and the McCulloch occupants of Barholm House would have been amused by and sympathetic to the smugglers and their trickery.

The Isle of Man is a short distance directly by sea from Barholm Castle and Creetown. In 1735, James, the 2nd Duke of Atholl, became Lord of the Isle of Man with a royal charter giving exemption from duty on many imports. Shipments of contraband rum, tea, brandy, wines and tobacco were sent from the Isle of Man to the Galloway coast on an industrial scale during the eighteenth century, with the Solway coast a favoured destination. Signals were sent to the smugglers' ships via fire beacons; the stone platform on the top of the tower at Barholm Castle would have been an ideal spot to place one. There was supposedly an underground passage from Barholm Castle down to nearby Kirkclaugh House, which would have been perfect for running contraband items from Dirk Hatterick's Cave on the shore of Wigtown Bay. However, no evidence for the tunnel's existence was ever found during the restoration of Barholm Castle. The story was unlikely to be true, given that the castle is built on unyielding bedrock and Kirkclaugh House is about a mile distant. We did hold our breath, however, when archaeological investigations were being carried out; although a secret tunnel would have been very exciting, it would have set back the restoration and cost us a great deal of additional resources to carry out any excavations.

The Lowland Clearances and the Galloway Levellers

At the time when the McCullochs were encouraging the growth of industry in and around Creetown, in the mid eighteenth century, the poor tenants and cottars of Galloway were suffering from the large-scale evictions known as the Lowland Clearances. These preceded the better-known Highland Clearances by more than half a century. By the early eighteenth century, the landowners of Galloway found that enclosing large herds of cattle before driving them to England for sale in drovers' markets was a great deal more lucrative than receiving rent in kind from poor tenant farmers who worked the land. A contemporary ballad by James Charters of Dalry shows the depths of anger felt by those dispossessed:

The Lamentation of the People of Galloway by the Pairking Lairds

> Among great men where shall ye find
> A godly man like Job,
> He made the widow's heart to sing,
> But our lairds made them sob.
> It is the duty of great men
> The poor folks to defend,
> But worldly interest moves our lairds,
> They mind another end.
>
> The lords and lairds they drive us out
> From mailings where we dwell,
> The poor man says 'Where shall we go?'
> The rich says 'Go to Hell.'
> These words they spoke in jest and mocks,
> But by their works we know,
> That if they have their herds and flocks,
> They care not where we go.
>
> Against the poor they still prevail
> With all their wicked works,
> And will enclose both moor and dale
> And turn corn fields to parks

(quoted in Aitchison and Cassell 2019, p. 35)

INTRODUCTION

The distress and suffering of the tenants led to a plan to destroy the dykes that the landowners had erected to enclose their profitable herds of cattle. A secret bond was drawn up in early 1724, pledging support to the 'Levellers', who, in organised gangs, set about pulling down, or levelling, the stone dykes enclosing the landowners' cattle. Levellers were active at nighttime in Wigtownshire and the Stewartry of Kirkcudbright, causing a great deal of damage and raising fear and alarm among landowners. In early June six troops of dragoons were stationed in the area, and the ringleaders were arrested. The Levellers carried on, sporadically, throughout the autumn, but by the end of the year the authorities had managed to quash the revolt. *The Lowland Clearances* by Peter Aitchison and Andrew Cassell gives a clear account of this relatively little-known episode in Scottish history which touched the lives of so many in Galloway. We do not know whether the dykes at Barholm Castle are original, or whether they were levelled at some point. It is clear that not all date from the same period; some are thicker, taller and more carefully constructed than others. There have been many changes in and around Barholm Castle over the centuries, both before and after its abandonment; it would be fascinating to have an analysis of what the dykes could tell us about the history of the land and its enclosure.

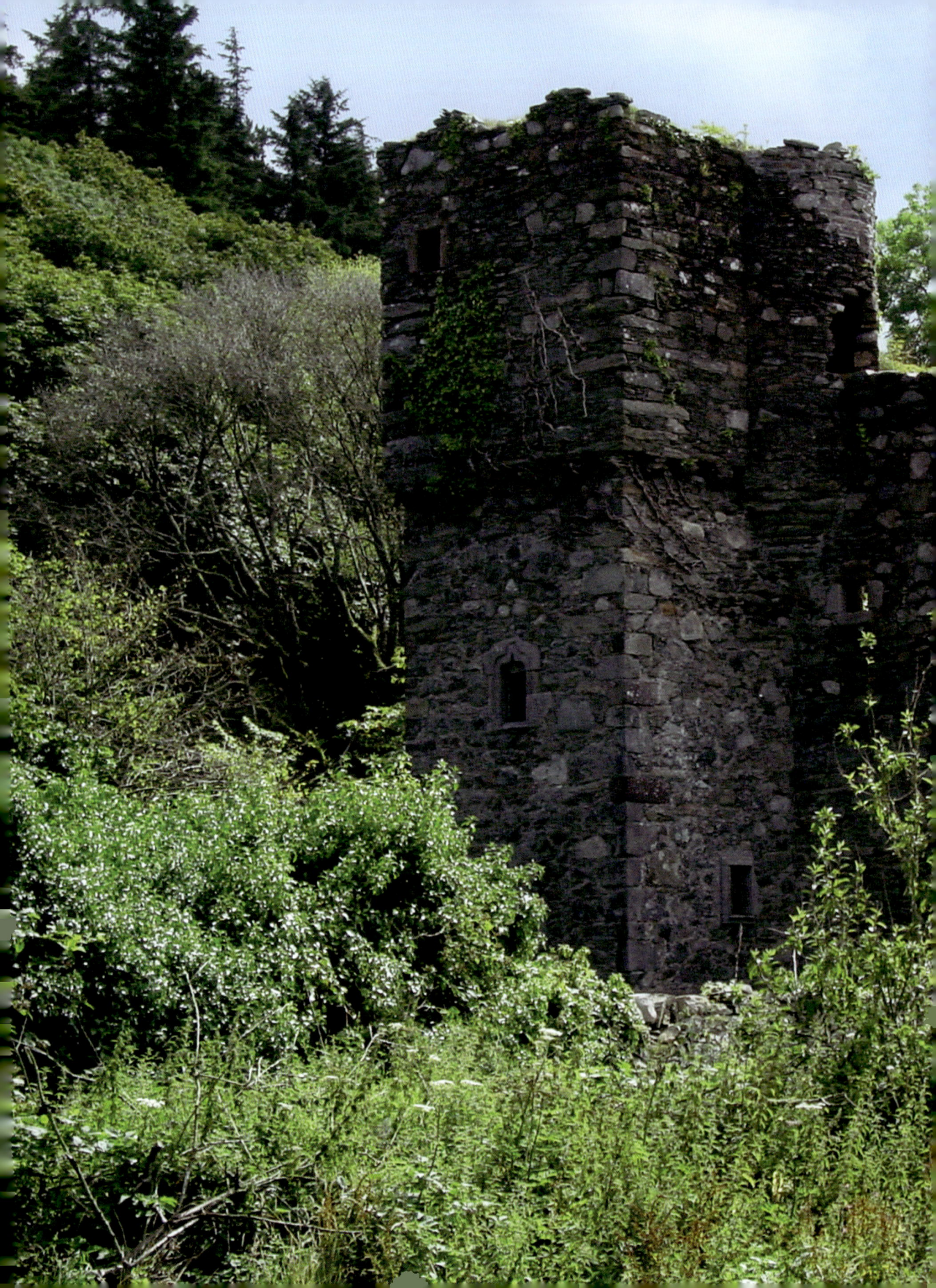

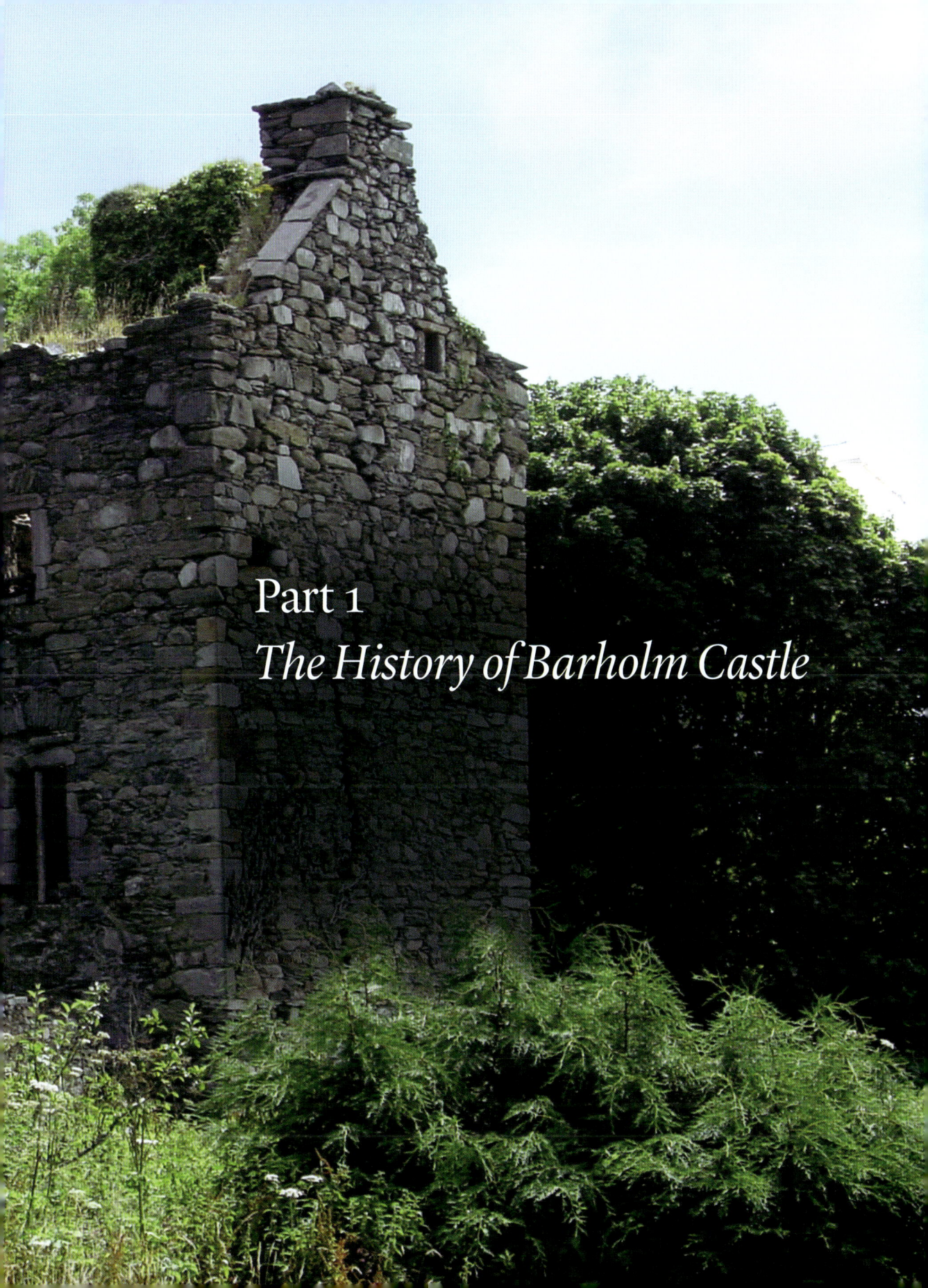

Part 1
The History of Barholm Castle

History is but a series of stories, and Barholm Castle has many of these. (One tale, told to me by former owner David Hannay, involved a pig from the farmyard making its way up the spiral staircase of the ruined tower and becoming stuck – sadly, there was not a happy ending.) Unravelling a narrative with a single thread has been impossible, but there are various stories that shine a light on what the past may have held. I have tried to impose some order on the somewhat chaotic, contradictory and possibly fanciful writings of the historians who have looked into Barholm Castle and the Barholm McCullochs.

The early history of Barholm Castle, including whether it existed before the sixteenth century, is mysterious, and its history after it became a ruin is lacking in any kind of detail. The meaning of the name *Barholm* is also uncertain. In Gaelic, *barr*, meaning top or height, is usually applied to hills of modest height. *Holm* in old Norse is a piece of land partly surrounded by streams or a stream. So 'a piece of land on a hill with a stream nearby' is plausible as an origin descriptor for *Barholm*, if not particularly special or specific. We do know something of the owners of the sixteenth and seventeenth centuries, and we know a lot about the ownership changes and plans in the second half of the twentieth century. We also know much about the building itself, although there are questions and mysteries remaining that may never be solved.

The Building

Barholm Castle, or Tower, is mentioned in most local history books and books about castles. While it is clear that many authors are simply repeating what has been previously published – the same sentences crop up again and again – some accounts are contradictory. Although most authors write in an authoritative tone, few give their sources. J.E. Russell, in his very thorough *History of Gatehouse and District*, claims that the tower had been standing since the late 1400s, but gives no details of its early owners or builders. The RCAHMS Inventory of 1914 claims that Barholm dates from the early years of the seventeenth century 'judging by the details'. However, when Professor Charles McKean visited Barholm in 2010, he was of the opinion that the building was much older than sixteenth century, because of the thickness of the ground-floor walls (about 8 feet thick). The thick ground-floor walls may indicate that this part of the building is of at least fifteenth-century origin, although it might simply be that the ground-floor walls are so robust because

THE HISTORY OF BARHOLM CASTLE

the tower is built on bedrock and foundations of any sort were thus impossible. (It is worth noting, however, that the walls of Ailsa Craig Castle, which dates from the sixteenth century, are only 1.5 feet thick and that tower is also built on bedrock.)

According to Alastair Maxwell-Irving,

> The earliest record of Barholm is in 1472, when Walter Porter of Blaiket wadset the £5 lands of Barholm, under reversion, to Donald Maclellan of Gelston. The Porters subsequently assigned their interest to David McCulloch, junior, in Conchieton, a grandson of James McCulloch of Cardoness, and it was presumably this David, now of 'Laggan-Mullen', who in 1541 sold Conchieton to one John McCulloch and his wife, the contract being signed 'at Barholm' – the first evidence of a building there. John McCulloch in Barholm finally redeemed the wadset in 1563, and a month later was granted a charter of Barholm by Thomas McCulloch of Cardoness. This was followed in 1570 by sasine being given to John in liferent, and to his son, James, in feu. Having thus secured their ownership, it was presumably at this time that the old 'house' was enlarged and improved, with the stair wing being added.
> (Maxwell-Irving 2014, p. 151)

Andrew Morton identifies another sixteenth-century mention of the name of McCulloch in connection with the lands of Barholm. This was in a contract dated 1 November 1528, concerning the redeeming of the 'Five pound land of Barholm' entered into between Thomas McClelland of Gelston and David McCulloch. The implication is that a building was extant at this point, or at least planned, but it was not until 1565, in a Charter dated 22 July, that John McCulloch (I) is described as 'of Barholm', a style of address that usually meant belonging to a house of that name.

The first recorded mention of Barholm as a building, apart from its appearance in Timothy Pont's map of *c.* 1590, is in the Rev. Andrew Symson's *A Large Description of Galloway* of 1684, where 'Barhoom' is listed in the appendix among the 'considerable houses' of the Stewartry of Kirkcudbright, which include Kardonesse Castle, Rusko Castle, Bardarach, Karsluith, Kassincary, Lairg Castle and Gairliss, the residence of the Earl of Galloway.

Barholm Castle presents as a very typical L-shaped sixteenth-century Scottish tower house, with a stair tower containing a spiral (or turnpike) staircase and one or two rooms on each of four storeys, accessed from the stair.

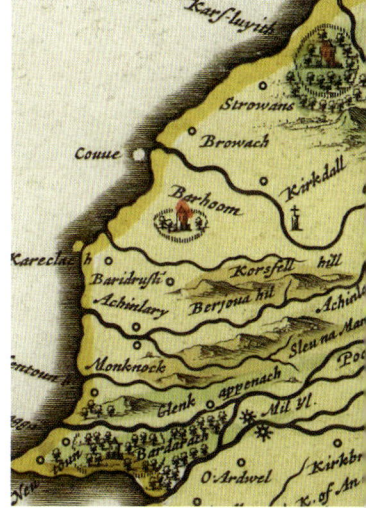

Timothy Pont's map of Galloway, *Gallovidae Pars Media*, made at the end of the sixteenth century. 'Barhoom' can be seen in between the castles of Kardeness (Cardoness) to the east and Karsluyith (Carsluith) to the west. Rusko Castle is depicted to the north-east, although Pont's maps do not lie directly on a north–south axis.

That is what MacGibbon and Ross assumed it to be when they visited *c.* 1880 and that is how it was described on the Canmore site prior to 2000. However, Alastair Maxwell-Irving was correct when he wrote: 'The architectural history of Barholm castle is more complex than it would first appear, and at times hard to explain' (Maxwell-Irving 2014, p. 151). He believed that the main house was originally a square tower with the entrance on the east side at first-floor level, and the stair tower wing was a sixteenth-century add-on. This explanation has plausibility, because of the (likely) precedent at nearby Carsluith Castle, and also because at first sight it seems obvious that the first-floor door in the east wall should have been the original main entrance, with tusking above to show that some form of structure, such as an external stair-case and/or balcony, had been built out – or intended to be built out – above it. This door would have overlooked the valley below, along which the road would have passed before the Robert Adam bridge was built across the Barholm Burn in the 1780s. Barholm Castle may not have been built for defence, but nonetheless it would have been useful for the inhabitants to see who was crossing the land in front of their property. Maxwell-Irving reports that Carsluith Castle, only a couple of miles away, had its stair tower added in 1568, and Andrew Wood, owner of Balbegno near Fettercairn, added a stair tower in 1569.

However, the team of archaeologists who studied the standing architecture of Barholm Castle from 2000 to 2004 disagree. Their report points out that:

> If the stair jamb was a later addition to the structure, it would be expected that there would be a recognisable scar where some of the existing masonry was removed to allow the jamb to be tied into the structure – such a feature is entirely absent. In addition to this, there is no evidence in the interior of the stair jamb that it has been forced into the existing tower, in short, the stair jamb is part of the original build.
> (Unpublished Archaeology report by Kirkdale Archaeology)

Discounting the idea that Barholm was originally a 'hall house' with later additions, the archaeologists instead proposed that the first phase of construction took place around 1580 – but give no evidence for their choice of date – at which point Barholm was built not to its final full elevation but as a three-storey tower house with a stair jamb of equivalent height. One argument against this theory is that shorter, squatter L-shaped laird's houses such as

THE HISTORY OF BARHOLM CASTLE

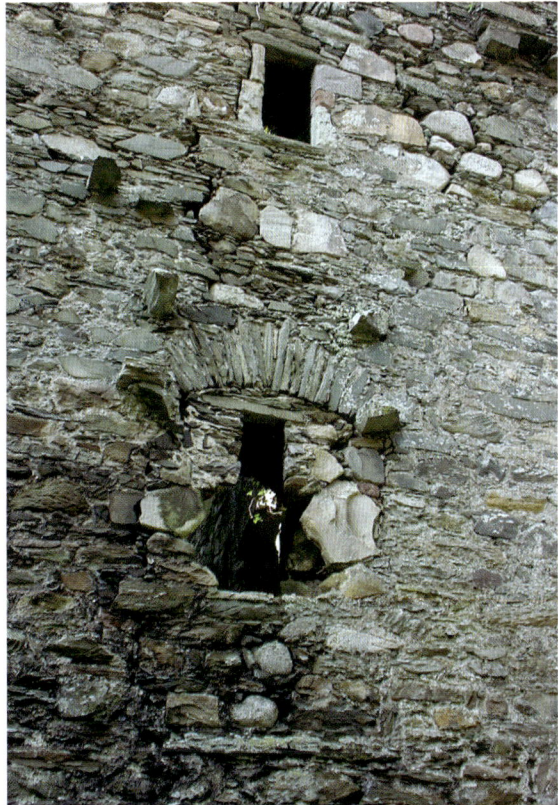 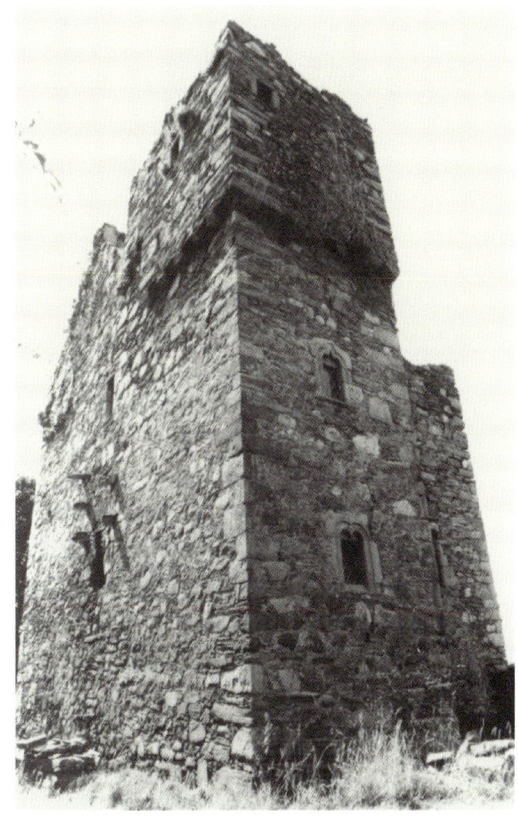

Left.
The infilled door at first-floor level on the east side can be clearly seen here.

Right.
The stair tower and east side before restoration.

this did not tend to be built until the seventeenth century, unless they were in towns – for example, Barscobe Castle near New Galloway. Sixteenth-century towers very much prioritised height and verticality. The ground floor, hall and second-floor chambers were, the archaeologists suggest, built much as they are now. Then in the second phase the height of the main roof of the building was altered in order to provide a further, higher level of private apartments, with the addition of the cap house and platform at the top. Heightening tower houses was not a common means of extending the accommodation: horizontal wings were usually added rather than roofs raised. The Castle of St John in Stranraer was altered upwards and outwards in the seventeenth century, but for a very specific purpose, to fit it for use as the town jail.

Although the absence of 'scarring' showing a join between the stair jamb and the wall on the east side might mean that both were built at once, we are inclined to question this as negative evidence. When we have had extensive repairs done on collapsed sections of the drystone dykes by experienced dykers, it is very difficult to see where the joins have once been. Equally, however, there are very clear signs in places where earlier dykers have filled in a section by crudely jamming stones into the gap. The amount of scarring depends upon the skill and diligence of the builders, and the extent to which the original building was deconstructed before the new part was added.

At Carsluith Castle, a few miles west of Barholm, the stair tower is said by Canmore to have been added in 1568, although Maxwell-Irving disputes this because the masonry is homogeneous – as indeed it is at Barholm. It is curious that he discounts similar evidence at Barholm in favour of the opposite conclusion. Carsluith, like Barholm, also has a first-floor doorway, but it leads to an external gallery, which allowed access from one room to another – a feature rarely seen in Scottish towers.

The archaeologists do not mention the insertion of the fireplace in the great hall. As Maxwell-Irving points out, 'Quite apart from the subtle differences in the masonry, the large chimney stack serving the hall fireplace is complete in itself above the level of the second-floor windows with its own quoins; at no point is its upper part keyed into the adjacent walls, implying that they were added, or rebuilt, later' (2014, p. 151). The clear evidence that the fireplace was added or rebuilt – perhaps by builders who were not assiduous or skilled in joining old and new masonry – implies a considerable investment in Barholm Castle. If, as Maxwell-Irving suggests, the castle was first built in 1541, probably by David McCulloch, one of his descendants must have decided to expand the accommodation by building upwards (the archaeological evidence is compelling here) and at the same time insert a huge chimney, a status symbol that would have signalled his wealth and power for miles around.

Castle historian Joachim Zeune claims that Barholm was altered around 1580 (the date the Kirkdale archaeologists suggest for the first phase of building) but offers no evidence (1992, p. 37). 1580 is not a likely date for any kind of building work, however, as James McCulloch II of Barholm was murdered in 1579 by John Brown of Carsluith, and his baby son, Thomas, was born posthumously in 1579. (James's brother John was also murdered, by Achilles John Maxwell, in 1613. The Browns and Maxwells were Catholics and the McCullochs dissenting Protestants, so these murders were probably crimes driven by religious disagreement.) There would have been no adult male McCulloch in a position to carry out building at Barholm until at least Thomas's majority, in 1601. Until then, his mother, Elizabeth Kirkpatrick, would have held the reins and managed the estate. Thomas is the most likely candidate for extending the building, as he lived until *c.* 1642, when he was in his sixties. If this was the case, the second phase of the building would most likely have been carried out in the first half of the seventeenth century. The work is less likely to have been done any time after that, as towers were beginning to become much less fashionable.

Like many of Scotland's castles, Barholm, as we have seen, has several mysterious aspects to its architecture. Whenever I encounter an unexplained architectural feature in an old building – and in the world of ancient Scottish castles there are many such examples – I think of a new house I watched being built in the Netherlands. It was on the route of my daily cycle ride, at the top of a dune where I always slowed down. A perfectly good 1950s house

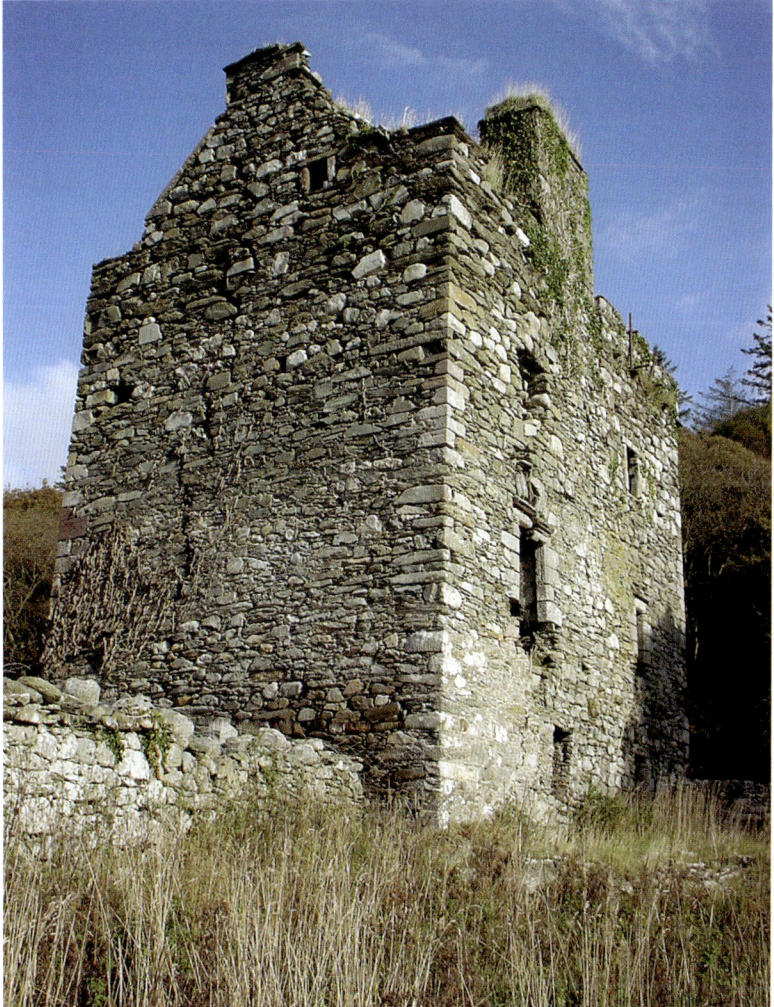

The south and west sides before restoration.

had been demolished, to make way for the newbuild, which was clearly being financed by someone with a great deal of money to spare. From the front, the house was built in French chateau style, with tall symmetrical windows flanked by shutters, a steeply pitched roof with blue slate tiles – unusual in the Netherlands – and a grand front door, set off by highly formal landscaping in the front garden. However, as I rounded the corner, I had the extraordinary sight of half a thatched cottage bolted on to the side of the house. What was the reason for such incongruity? I can only guess, but I think that the owner, who clearly had excessive money to spend, was unable to choose between the cosy vernacular, as seen in the large number of thatched mansions masquerading as chocolate-box cottages in the neighbourhood, and the more austere and regular chateau style, which was much less common but certainly up and coming. I imagine a conversation with the architect which ended along the lines of 'Well, why not just have both, if that's what you want?' In the same vein, I imagine a sixteenth-century Scottish builder–architect saying to a difficult client (architects always claim to have difficult clients), 'Nae bother, sir, if ye want another wee turret wi' an outlook tower on top up there and some big pointy spouties along the side, we'll see that it's done for ye.'

The door and stair tower windows

The door of Barholm Castle, with its elaborate mouldings, has given rise to much speculation as to their significance. The Romanesque segmented arch is surmounted by a rope tied loosely at each end, a motif also seen on the window frames of Kenmure Castle, near New Galloway in Kirkcudbrightshire. Rope moulding can also be seen on the Victorian courthouse in Dumfries and on the Victorian door of Drum Castle in Aberdeenshire; this latter example is remarkably similar to Barholm's, minus the heads and the beast. The two heads on the Barholm door mouldings are said to be 'death's heads', and the creature just off-centre at the top – which has been carved out of a different stone, redder than the other segments – is described variously as a salamander, a grotesque animal and a doglike creature. Richard Aytoun, on his 1813 tour of Dumfries and Galloway with the artist William Daniell, observed 'Over the doorway of Barholm is some sculpture representing a folded serpent, death's heads, with several non-descript monsters' (quoted in Macleod 1988, p. 139). Given that the doorways of sixteenth-century lairds' houses are almost without exception plainly rectangular, it seems likely that this decorated doorway was recycled from another, grander building,

THE HISTORY OF BARHOLM CASTLE

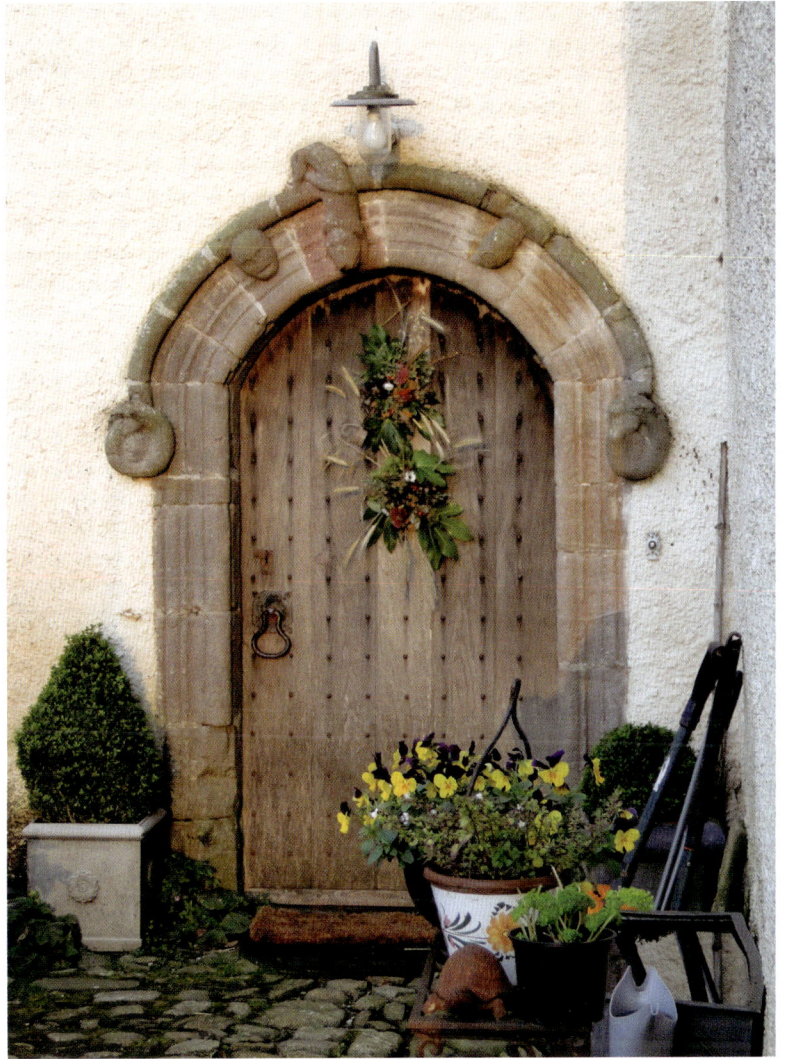

Left.
The elaborate door of Barholm Castle.

Below.
The Victorian door of Drum Castle in Aberdeenshire.

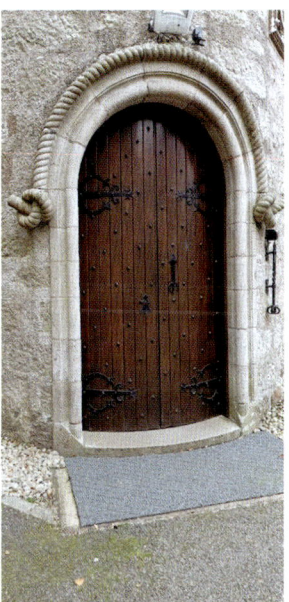

probably ecclesiastical. The two windows on the north side of the stair tower are also more elaborate than would be expected.

The oak door was made especially for us during the restoration. Its construction, like all of the internal doors, consists of a double layer of timber forming outer and inner boards set in opposite directions. They are held together by sturdy iron nails, the stud heads of which can be seen on the vertically placed outer planks, and the bent nail ends on the other side. The internal doors all have a modern intumescent strip around the edges, as a fire prevention measure. There was not a scrap of wood left in the ruin of

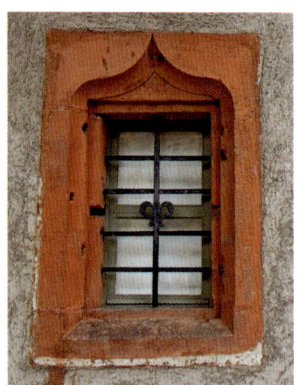

Ogee window in the Brunnenhalle at Heidelberg Castle in Germany.

Barholm Castle; it would all have rotted away centuries ago. A fellow castle restorer in Ayrshire told us that he had had to replace his softwood window frames three times in twenty-five years because the wood had rotted. Oak was therefore our wood of choice for the timber window surrounds and the door, despite its high cost. At Barholm Castle, situated on the coast nearly 400 feet above the sea, the onslaught from wind and weather would doubtless have led to the fast decay of any exposed softwood.

Orroland House, near Dundrennan Abbey, probably built in the seventeenth century, has evidence pointing to an earlier core, and has a window very similar to the double-arched one on the stair tower at Barholm Castle. These elaborately carved windows may possibly have been taken from local abbeys after they fell into ruin at the beginning of the seventeenth century. Maybole Castle in Ayrshire also has a window similar to the Gothic ogee one on Barholm's stair. Perhaps all were the hand of the same mason, even up-cycled from the same source. Heidelburg Castle in Germany has a very similar ogee window in its Brunnenhalle; clearly, it was a popular design, although one not frequently executed.

The construction of Barholm Castle

Barholm was almost certainly built without the benefit of measured drawings and plans, since these were not in common use until the eighteenth century. Indeed, it is likely that the very first plan of Barholm Castle was drawn by MacGibbon and Ross in around 1890 for their virtuoso five-volume survey of the castellated and domestic architecture of Scotland, accompanied by a sketch.

Barholm Castle was built using the hard and unyielding greywacke local stone. Some of the building material would be field gatherings, as used for dykes across Galloway. Other stone would have come from quarrying the rock behind the east side of the tower, where there is clear evidence that the stone has been cut away, leaving a sheer rock face bordering the ravine. This, of course, would make access to the tower all the more difficult for unwanted visitors. Wood was in relatively short supply in sixteenth-century Scotland, as there were far fewer trees than there are today. The width of a tower such as Barholm was determined by the ease with which long straight tree boles for the ceiling beams could be sourced. Most of these were imported to Scotland from Scandinavia and the Baltic countries.

Quirks in the building are everywhere. The 'spoutie' above the front door

THE HISTORY OF BARHOLM CASTLE

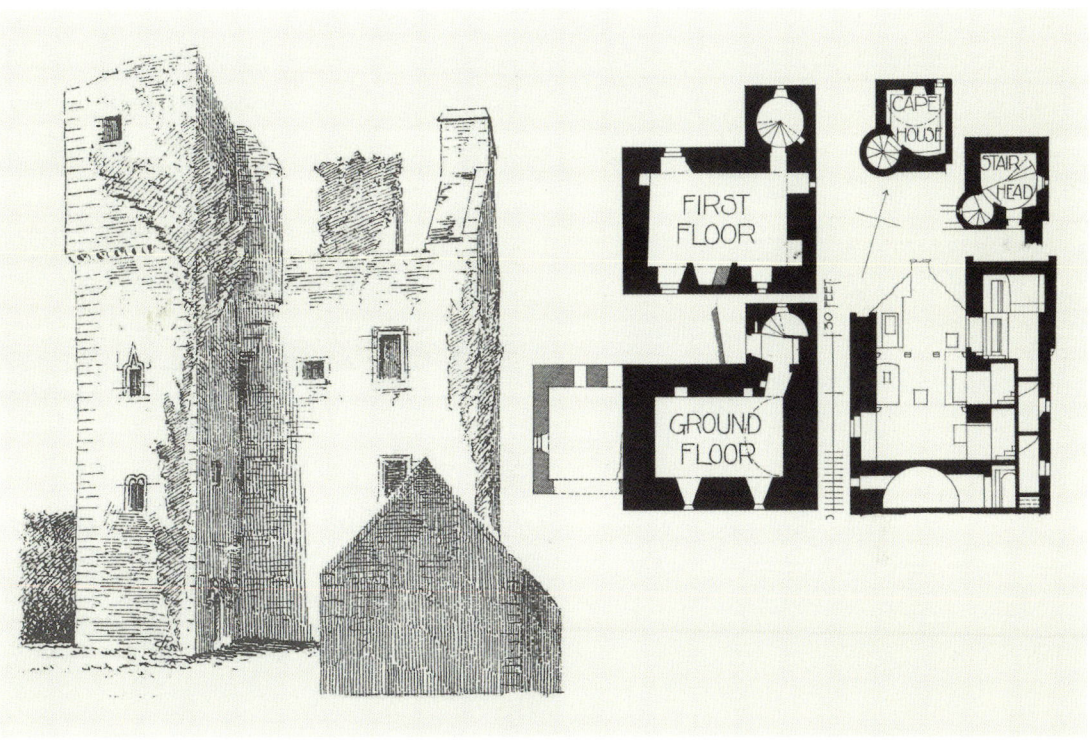

Above left.
Sketch of Barholm Castle by MacGibbon and Ross (1889).

Above right.
Plans and section of Barholm Castle from the RCAHMS Inventory of 1911.

pours a jet of rainwater straight down on anyone unfortunate enough to be leaving or entering the building in wet weather. That must surely have been just as much of an inconvenience to the sixteenth-century inhabitants as it is to us today. I expect they cursed the builders then as we do now. Vertical drainpipes, made of cast iron, were not in common usage until the seventeenth century. They can be seen on the walls of Craigcaffie Castle, the Isle of Whithorn Castle and Machermore Castle, indications of work carried out in the eighteenth and nineteenth centuries to upgrade the buildings for more demanding and sophisticated inhabitants. Larger windows would have been installed at the same time – but not at Barholm.

The building contract and specification for Partick Castle, dated 1611, demonstrates the lengths a careful owner would go to in order to make sure every little detail was signalled to the builder: 'In ye gabill of ye Jame of sufficient forme and quantitie as becomes by the length of ye saids saxtein fuittis Ane turnpyik to be biggit and raisit be it selff at ye northeist nuk of ye mainhous of nien or ten futis wide wtin ye walls.' ('In the gable of the jamb of sufficient form and quantity as becomes by the length of the said sixteen foot, a turnpike [i.e. spiral staircase] to be built and raised by itself at the north-east

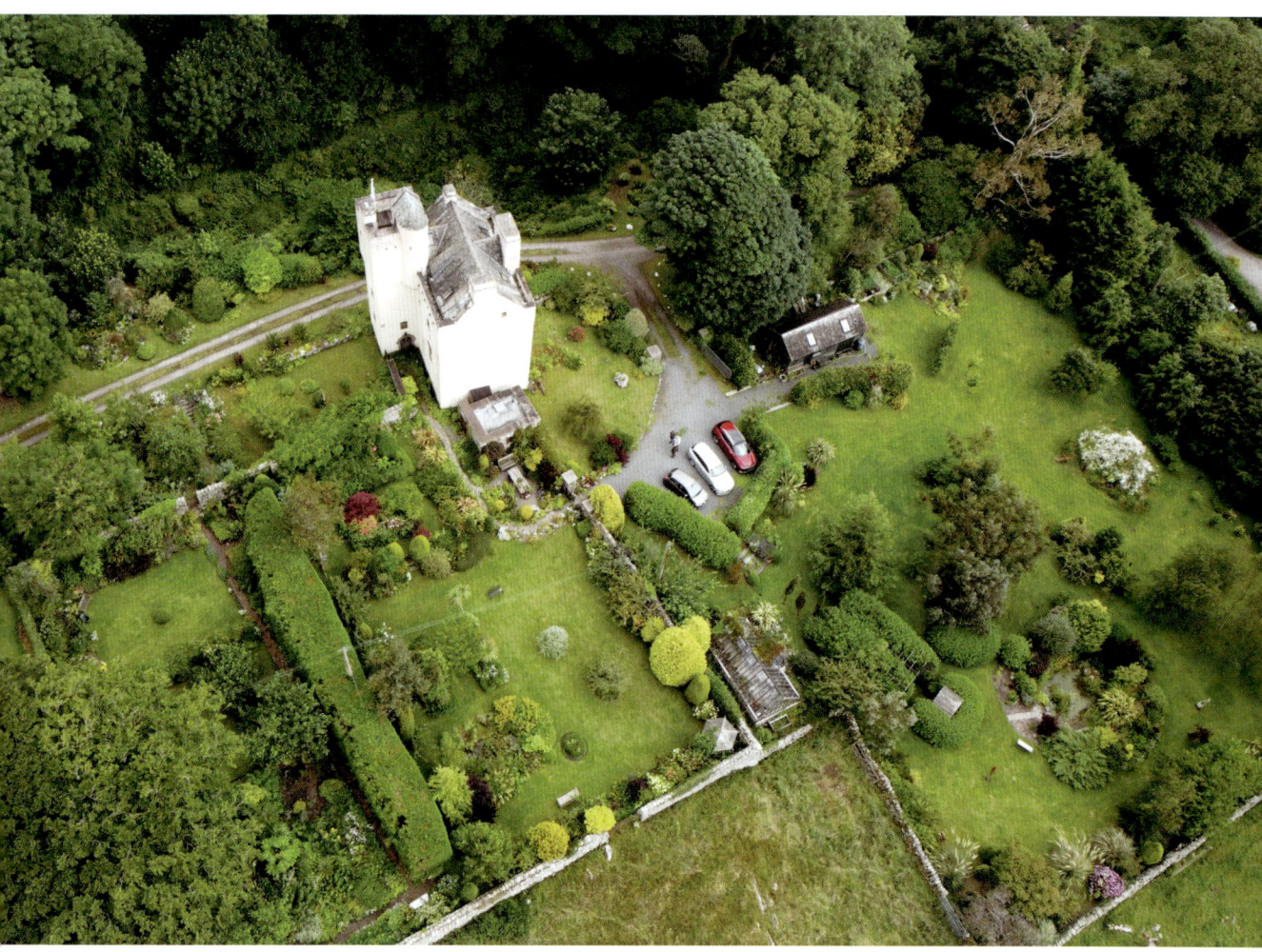

Aerial view of Barholm Castle taken in 2024. It can be seen from this photograph that the tower is not square with the enclosure walls.

corner of the main house of nine or ten feet wide within the walls.') The internal width of the spacious turnpike staircase in Barholm Castle is about 8 feet, so Partick Castle's would have been a very commodious size. Unfortunately, it was demolished in the 1830s; its remains now lie beneath a Tesco supermarket in Glasgow.

It was only during the period of Covid lockdown, when we had paths constructed up the far side of the steep ravine behind the tower, that we were properly able to study its east side. Before that, we could only stand a few feet adjacent to the wall and look sharply upwards; to step further back would be to risk tumbling down the sheer drop where the quarrying had been carried out. I was struck by the unevenness of the corbels, like crooked gappy teeth sticking down from the upper course of the stair tower. These were constructed from local stone and it must have been horribly difficult to source exactly the right size and shape of stone to match and be positioned in the right place, 20 feet above ground level. It is doubtless why the holes for the ceiling beams, and thus the beams themselves, are not placed at completely regular intervals. Working with large pieces of random stone at a great height would militate against precision. The great palaces and houses of the titled and wealthy tend to be more symmetrical in execution; they had the resources to pay for a large skilled workforce, whereas the McCullochs doubtless had to make do with whoever was to hand locally. In fact, most of the angles throughout Barholm are skew-whiff when measured, or 'eye-sweet' as our architect called it. The archaeologists pointed out that the walled enclosure of the walled garden is not square with the walls of the castle, as can be seen in the aerial photograph. This may mean, they speculate, that it was built after the first phase of tower building, although on the Ainslie map of 1782 (see page 67) it is depicted – probably erroneously – as square. The old tower of Cawdor Castle near Inverness is also set at an angle within the walls of the later additions that surround it.

Water?

Much is unclear about life at Barholm in the early days. One question that we are often asked is 'Where is the well?' 'Nowhere', seems to be the answer. The tower is built on bedrock and there is no sign in the surrounding area of anywhere that might have been dug out as a well. Nor is there a river or stream in the immediate vicinity, although the fast-flowing Kirkdale Burn is within walking distance, down a steep hill. Some mediaeval castles in Europe used cisterns which collect rainwater, either by a system of filtering or in open tanks. I have placed open faux-lead containers under some of the drainage spouties, of which there are fourteen altogether, in order to catch rainwater for my plant tubs. It is feasible that the original occupants also used more and larger cistern containers to collect rainwater, which would be used for cooking, ale-making and washing. The current annual rainfall for Dumfries and Galloway is over 1,000 mm per year, spread fairly evenly across the seasons. It may have been less in the sixteenth century, but would still have been sufficient to meet the needs of a family and their servants in less hygienic times, topped up by water from the Kirkdale Burn when necessary.

The road

Barholm Castle sits high above the A75 trunk road, which runs between the castle and the sea. The A75 links Stranraer in the far west of Galloway to Gretna, on the Scottish–English border. We cannot see the road from the garden, and can see only a tiny stretch of it from the viewing platform at the top of the tower, so we are well shielded from traffic noise. The A75 has not always followed the same route, however; there have been many changes over the decades as villages and towns have been bypassed and meandering stretches straightened out. The road was only officially named and numbered in 1923, as part of the Great Britain road numbering scheme. In 1787, Kirkdale Bridge, spanning the Kirkdale Burn, was built for Sir Samuel Hannay, based on designs drawn up by Robert Adam, doubtless to make road access easier to the newly built Kirkdale House and its estate. Although the Kirkdale Burn is little more than a stream, because of the steepness of the gorge through which it runs, the bridge is high and triple-spanned, with the central arch spanning 6 metres. Before then, the road would have run *behind* Barholm Castle, parallel with the east side. We can look down from the great hall window, which was once a door, and see the line of the former road running along the side of the ravine. There is a retaining wall of local stone, now overgrown with ferns, with a flat path in front of it, about one cart's width. This route would make a great deal of sense in terms of the position of Barholm Castle; the occupants would have been able to keep watch over the road and gather intelligence about who was coming and going across McCulloch land.

The McCullochs of Barholm

The history of any house is also the history of its owners. Although we have no personal connection with any branch of the McCulloch family, we know a great deal about the history of the Barholm McCullochs. This is partly because modern-day McCullochs have frequently turned up here, many bringing us their family history research and pieces of information that help us piece together the story of the castle and the family. Michael MacLeod's book *Creetown's Industrial Era* is very helpful on the McCullochs since 1770, and Walter McCulloch's unpublished manuscript *History of the McCullochs* tells us that the McCullochs were rather a wild and passionate lot. They espoused the Reformation and Covenanting causes, a dangerous thing to do in the seventeenth century. Two of them were murdered by Catholic

THE HISTORY OF BARHOLM CASTLE

neighbours, one was executed for his religious beliefs, and one became an outlaw after leading Galloway troops in battle. Another McCulloch was wanted for the murder of a neighbour, although the motive there seemed to be clan loyalty and greed. McCullochs had a fearsome reputation in Galloway and beyond, as evidenced by the 'Curse of the McCulloch':

A Manxman's Prayer

Keep me, my good corn, and my sheep and my bullocks
From Satan, from Sin, and those thievish McCullochs.

There is disagreement about the dates when the McCullochs took over Barholm. Various branches of the McCulloch family held a number of lands in the Wigtown area, with the main branch at Myrton, near Monreith House, which still stands. The original owners seem to have been the McClellands:

> Barholm was the third cadet house in the barony of Cardoness . . . in the original parish of Kirkdale, and later in Kirkmabreck parish. It first seems to have come into McCulloch hands when David McCulloch, younger son of James McCulloch of Cardoness, received Barholm and Conquhieton by Crown charter as a military tenure in 1507: David McCulloch was referred to as 'of Barholm' in 1536. David McCulloch had three sons: James McCulloch of Bardristane, John McCulloch of Conquhieton and Alexander McCulloch who became the tenant of Ardwall . . . James McCulloch was infeft [symbolically in possession] in 1522 in the three merk lands of Bardristane, probably a gift from his father, and probably Barholm at the same time, although he never took sasine [possession] on Barholm, which was still under wadset [mortgage]. (Russell 2000, p. 25)

Russell goes on:

> John McCulloch I succeeded to the Barholm estate and redeemed the wadset on Barholm, possibly by raising a wadset on Bardristane in 1563, and received the charter for Barholm from his feudal superior, Thomas McCulloch of Cardoness . . . he probably remodelled Barholm tower house, which had been standing since the late 1400s, and in his third deed of 1563 he issued a sasine

'proprus manibus' in favour of William Cullintnick and his wife Marion Muir, of the 3 merk land of Barholm of old extent, perhaps to pay for this work. In April 1570, Thomas McCulloch of Cardoness granted a charter of resignation of Bardistrane to John McCulloch I of Barholm in liferent, and to his son and heir James McCulloch in fee: presumably his debts raised by the work on Barholm castle were liquidated . . . James McCulloch II of Barholm was a minor when he inherited Barholm and Bardistrane and it was not until 1575 that he was served heir to his father.
(Russell 2000, p. 26)

Russell is convincing in his McCulloch details, but his assertion that Barholm Tower had been standing since the 1400s is unsupported by evidence and highly unlikely, given the building history of Galloway towers. There may, however, have been an earlier building on the site, given that it was common practice to build where there had already been some structure in existence.

Morton disagrees over the dates and some details:

The first mention of the name of McCulloch in connection with the lands of Barholm was in a contract dated 1st November, 1528, anent the redeeming of the 'Five pound land of Barholm' entered into between Thomas McClelland of Gelston and David McCulloch, who appears to have been the younger son of David, second son of James McCulloch of Cardoness; and then in a Charter dated 22nd July 1565, is styled as of Barholm.
(Morton 1925–6, p. 233)

There is not space here to delve into every detail of McCulloch history; Walter McCulloch's manuscript, now available online, is exhaustive in its treatment of the Barholm branch and many others besides and should be consulted for the minutiae of family history. However, he cautions:

The account of this branch of the family must, unfortunately, remain incomplete. For reasons best known to themselves, successive representatives of the family have firmly and repeatedly refused to allow the family documents and papers to be examined. It is believed, and from what is known of the family, the belief seems well founded, that Barholm [House], prior to its recent demolition [in 1959], contained a very large number of old papers

concerning not only the history of this, but also many other
Galloway families which could well be of local antiquarian interest.
(W. McCulloch 1964, p. 155)

The family tree, or 'Pedigree Chart of the Family of McCulloch of Barholm', as set out by Walter, begins with David McCulloch of Barholm, son of James McCulloch of Cardiness. David married Margaret Menzies, and their older son, James McCulloch of Bardristane (a mile west of Barholm Castle), inherited on David's death in 1541. James married Margaret Gordoun. He died in 1562 or 1563 and their oldest son, John McCulloch (I) of Barholm, took over. John McCulloch I received a charter of Barholm from his superior and kinsman, Thomas McCulloch of Cardiness, in 1563. If John Knox did indeed shelter at Barholm Castle in 1566, as the story goes (see page 52), then he was sheltered by John McCulloch I. However, John had been the laird of Barholm for only nine years when he died in 1571; his son James was a minor at succession.

James, too, had only a short time as laird, but he did marry and have two children, Thomas and Janet. Tragically, on 17 April 1579 he was murdered by his neighbours, John Broun of Carsluith and John Bek in Kirkbryde. Walter McCulloch tells the story: 'His widow, Elizabeth Kilpatrick, his bairns and kinsmen, raised Letters against Carsluith, who found Laird Maxwell as surety that he would underlay the law, that is, stand their trial, for the crime. But Carsluith went into hiding and Maxwell was amerced [fined] £40 for not producing him' (W. McCulloch 1964, p. 162). James's brother, John, was also murdered several decades later in 1613, by Achilles John Maxwell. These were dangerous times.

James's son, Thomas McCulloch, was a posthumous child. On 12 March 1601, when he came of age, he became heir to his father in the three-merk land of Bardristane. He married Mary McKie, possibly the daughter of Alexander McKie of Broach, who held a wadset (mortgage) of 1,000 merks on a part of Barholm. Thomas's date of death is not clear, but is likely to have been around 1642, according to Walter McCulloch. Thomas, as has already been shown, is the most likely candidate for carrying out the additional building at Barholm, which certainly consisted of adding the large chimney for the great hall and building up the long gallery on the third floor, and possibly adding the stair tower. The archaeologists' report claims that John Knox's room (see page 96) was added in the second phase of building, after 1580, but if Knox was there in 1566, then Thomas cannot have been the one who commissioned the work; it would have to have been his grandfather, John McCulloch (I).

Covenanting connections

In 1662, after the restoration of the monarchy, Scots were required to renounce the 1638 National Covenant, a bond in which they had pledged to maintain their support for a Presbyterian Church in Scotland and the primacy of its leaders in religious affairs. Signatories of the Covenant were opposed to interference from the Stuart kings in the affairs of the Kirk and rejected the belief in the Divine Right of kings to be spiritual as well as secular leaders. King Charles I had introduced the Book of Common Prayer to Scotland in 1637, to the fury and resentment of much of the populace. He declared that opposition to the new liturgy would be treason, and thus came about the Covenant. When Charles was restored to the throne in 1660 a period of very severe repression followed. Ministers with Covenanting sympathies were 'outed' from their churches by the authorities and had to leave their parishes. Many continued to preach at 'conventicles' in the open air or in barns and houses. This became an offence punishable by death. Citizens who did not attend their local churches (which were now in the charge of Episcopalian 'curates') could be heavily fined, and such offenders were regarded as rebels, who could be held for questioning, even under torture. They could be asked to take various oaths, which not only declared loyalty to the king, but also accepted him as head of the Church. Failure to take such an oath could result in summary execution by the muskets of the dragoons, who were scouring the districts looking for rebels. Galloway was home to many Covenanters during the 'killing times' leading up to the Glorious Revolution of 1688, when the House of Stuart was overthrown.

The McCullochs warmly espoused the cause of the Covenanters, and for his nonconformity Thomas's son, Major John McCulloch II of Barholm, was fined £800 in 1660. Soldiers were quartered on him for thirty days at a time: he had not only to keep them but to pay them. The Major subsequently took a leading part in the insurrection known as the 'Pentland Rising' and was taken prisoner at the Battle of Rullion Green in Midlothian in 1666. The Privy Council ordained his head to be cut off and sent to Kirkcudbright to be stuck on the market cross for exposure, his right hand to Lanark, where he had taken the Covenant with uplifted hands, and his body buried like a traitor. However, he was spared the final indignity: Major McCulloch was buried in Greyfriars Churchyard, Edinburgh, where his epitaph reads, 'Major John McKoolo, west countryman, executed.' According to Walter McCulloch, '[It] seems probable that John McCulloch had served abroad in the continental wars, and, at the time of the [Pentland] rising, had been in retirement at Barholm, for, in 1666 he is described by Woodrow as a much respected and reverend old gentleman' (1964, p. 163).

After the Major's execution, on 7 December 1666, property belonging to his son Henry, who was not concerned in the rising, was seized, and Henry was imprisoned for a year. He carried on defending the Covenanting cause after his father's death. On 3 August 1676, he was among those ordered to be denounced for harbouring, resetting (sheltering an outlaw) and speaking with proscribed persons. He commanded the Galloway Covenanters against government troops at Bothwell Bridge in Lanarkshire on 22 June 1679, where they were defeated after a hard-fought battle. Henry escaped capture but was tried in his absence and, with many others, ordered to be executed when apprehended. The sentence, however, was never carried out. The Barholm estate continued under forfeiture until the Revolution of 1688–9, when it was restored.

Andrew McCulloch in his book *Galloway* looked at the wider McCulloch clan and noted that 'Other families of the same name [i.e. McCulloch] . . . continued to be small landholders or "bonnet lairds" in Galloway, the only exception being the McCullochs of Barholm who were more substantial landowners, though at this time [the 1600s] they were under a cloud in consequence of their adherence to the Covenant' (2000, p. 398).

In 1696 John McCulloch III inherited the title of Barholm from his grandfather, Major John McCulloch II, and in September of that year he inherited from his mother, Isobel McDowell, the estates of Pybill and Claughreid. He had married Jean Gordon, only daughter of John Gordon of Culvennan, from whom came additional lands including Ferrytown of Cree and its mill. He died in 1747 and John McCulloch IV of Barholm succeeded him.

By this time, tower houses were unfashionable and a move to an Adam house was already being planned: the heyday of tower house building was from 1560 to 1610, although the earliest towers date from the thirteenth century. By the middle of the seventeenth century, the tower house had gone out of fashion and was no longer being constructed as an architectural form:

> Moralising upon it in his own case, the first Earl of Strathmore (1677), with his own Castle Lyon, now Huntly, with Glamis in his mind, could comment that 'such houses truly are worn quyt out of fashione, as feuds are . . . the country being generally more civilised than it was in ancient times.' It was no longer necessary for a man to 'make himself a prisoner' in such a place. 'There is

no man,' he continues, 'more against these [sic] old fashion of tours and castles than I am.' (Scott-Moncrieff 1938, p. 76)

However, Barholm McCullochs were still living at Barholm Castle in 1771. It is likely that the old tower was used as a cadet house for the Younger McCullochs (i.e. the heirs) from the time when the McCullochs moved to Balhassie/Balhazy in Creetown, probably around the early eighteenth century when John McCulloch III married Jean Gordon of Culvennan. John McCulloch the Younger (IV) was actively making plans to move to Creetown himself. We know this from an advertisement he placed in the *Edinburgh Evening Courant* of 11 September 1771:

> that he is resolved to give the greatest encouragement to all merchants, tradesmen, and others who may incline to build and settle there ... There is a piece of ground nigh the town [i.e. Creetown], of considerable extent, which will be let in parcels, larger or smaller as feuars may incline, any time betwixt this and Whitsunday next, by said John McCulloch younger of Barholm, to whom any person choosing to take less or more of that ground may apply, at his house of Barholm, near Gatehouse of Fleet, either personally or by letter.

Disaster strikes the McCullochs

In 1747 John McCulloch IV inherited the Barholm title and estates, including Anwoth, which his mother Jean Gordon had acquired. He was known by his family to be profligate with money and both his sister, Isabel, and John McCulloch III, had written entails into their wills to curb his inheritance. John McCulloch IV executed a number of legal actions to try to overturn these and gain more control, but only succeeded in racking up debt. In 1773, disaster struck the McCullochs of Barholm. John McCulloch's creditors – among whom were the brothers Alexander and Nathaniel Gordon of Kirkcowan – petitioned for Parliament to pass an act forcing the sale of land to clear his debts. William Hannay of Kirkdale, Barholm's nearest neighbour, was appointed Judicial Factor and took advantage of his position to sell himself the lands of Barholm, Broach, Claughreid and Cambret in the public auction of 1775. The Hannays continued to own the lands around Barholm from then until 1960, when they were sold to the Hoggs.

THE HISTORY OF BARHOLM CASTLE

John McCulloch V, son of John McCulloch IV, was born around 1740 and inherited Barholm when his father died in 1778. It was by this time a much-diminished estate. Tens of thousands of acres of arable land and estates had been sold in the public roup (auction) of April 1775, mainly around Creetown and including Barholm Castle. It is telling that the sales particulars noted that the 'lands are very low rented and capable of great improvement'. However, the advertisement went on:

> The Lands of Barholme and Bardristane have a neat dwelling house and commodious office-houses, delightfully situated on the mouth of the Bay of Wigtown, commanding a view of England and Ireland, and the Isle of Man; and upon the lands of Barholme there is a very thriving wood, mostly of ash and oak, to the extent of between 40 and 50 acres.

If the 'neat dwelling house' was Barholm Castle, William Hannay clearly decided that it was not worth having, except as a useful adjunct to the farm buildings. According to Canmore, Barholm farmhouse, which lies very close to Barholm Castle, was built in 1797, almost certainly by William Hannay after he took over the land. From then on, Barholm Castle was used as a farm store. Fortunately, the tower was not dismantled for building materials for the farmhouse, as sometimes happened. Three things probably helped: the abundance of natural building material that was already freely available in the area; the usefulness of a building already in situ for storage of farm materials, even if the upper-floor roofs and ceilings disintegrated over time; and the sheer work involved in dismantling a tower with 8-foot-thick walls at its base.

The move to Barholm House

John McCulloch V was not long deterred by the loss of much of the family land. A quarter of a century earlier, in 1754, his father, John McCulloch IV, had commissioned William Adam to design a house for his estate in Creetown, six miles west of where Barholm Castle stands. This was included as an illustration in Adam's collection of grand plans and elevations, *Vitruvius Scoticus* (plate 94), though it was never executed. In 1787–8, William Adam's son Robert was commissioned to provide designs for a revised scheme. The new scheme shared some similarities with the 1754 design, comprising a

Design of a House for John McCulloch Esqr of Barholm in GALLOWAY

P.94

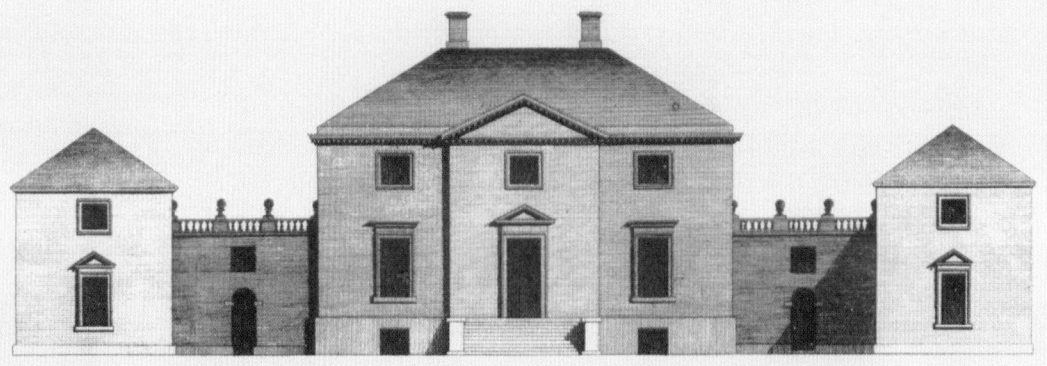

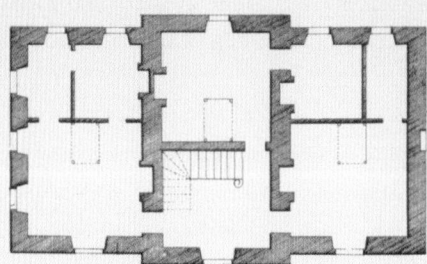

Plan of the Attic Story

Principal Floor

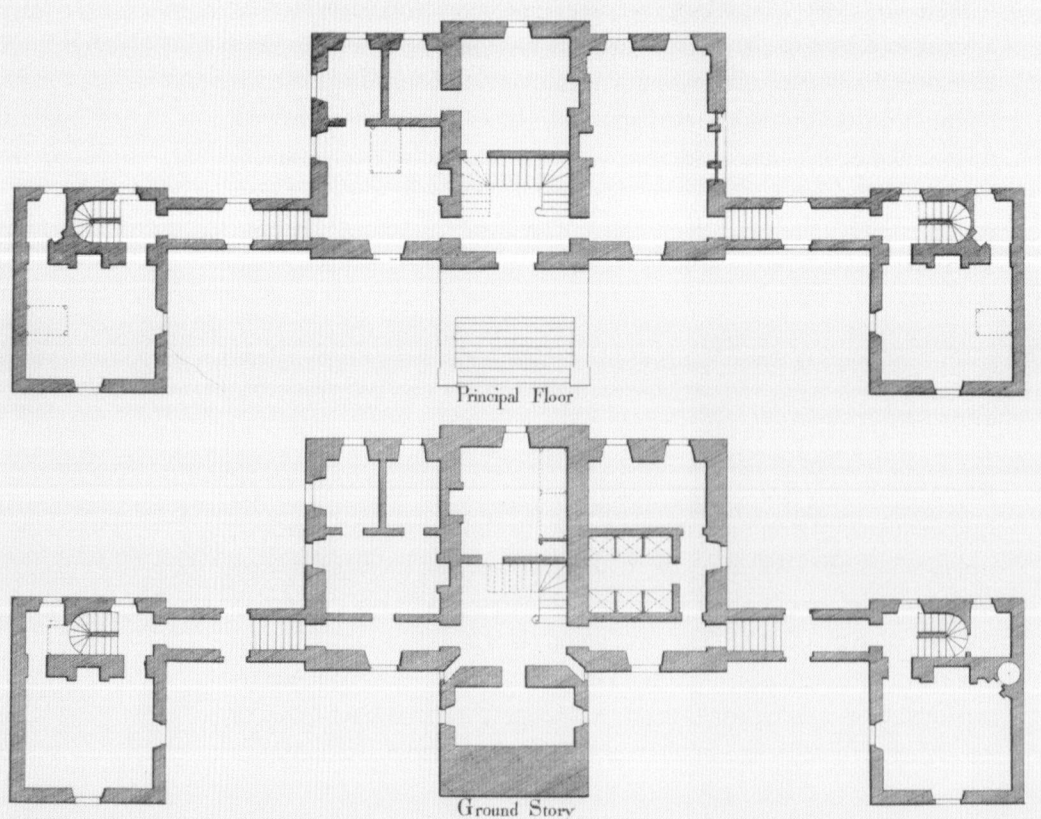

Ground Story

Adam Archt.

Mazell Sculpt.

central three-bay block with flanking wings that terminate in single-bay pavilions. Construction of Barholm House was finished in 1792.

At the same time as John McCulloch V planned his new house, Sir Samuel Hannay had been building an elegant house constructed of shining Creetown granite at Kirkdale, which he had also commissioned from Robert Adam, only half a mile from Barholm Castle and easily visible from there. It may be that the old tower house, now owned by the Hannay family, was kept within the line of sight of Kirkdale intentionally, as a symbol of ancient lineage acquired. Sir Samuel Hannay was the son of the William Hannay who had taken possession of so much of the McCulloch land; there was doubtless rivalry between him and John McCulloch V. The *Second Statistical Account* of 1840 described Sir Samuel Hannay's house thus:

> Kirkdale House, the seat of Miss Hannay, is a magnificent building of the Grecian order of architecture, from the design of Mr Adam. It is wholly built of beautiful polished granite of exquisite workmanship; and, until the late repairs upon the princely mansion of Mr Murray of Cally, M.P., it was without a rival in Galloway. It is stated in the old Statistical Account that the Bishop of Derry, on seeing the house of the late Sir S. Hannay, was so much charmed with the stone, that he immediately contracted with the superintendent of that work for the building of the spires of two churches in his diocese, which were all to be executed with this stone.

Kirkdale House still stands and is in good condition. It was divided into flats in the 1960s and is still owned by the Hannay family.

Barholm House, although elegant in a typically Adam style, was built of local stone (not granite, despite the fine quarry nearby in Creetown) and does not have the imposing panache and swagger of Kirkdale. It is a lesser house, which probably piqued John McCulloch V's pride and allowed Samuel Hannay to feel justifiably superior.

Between leaving Barholm Castle at the latest in 1775 – and probably well before this – and moving into Barholm House in Creetown in 1792, the McCullochs lived in a house called Balhazy, or Balhassie, just outside Creetown, on what is now Station Road. There is no longer any trace of the house; it may be that it was purchased by the railway company in the mid nineteenth century and demolished to make way for the track. Perhaps John McCulloch V was not sorry to leave Barholm Castle. The sheer inconvenience of living in a small tower with only five vertically arranged rooms, few windows and

Opposite.
William Adam's 1754 plans for Barholm House, illustrated in *Vitruvius Scoticus* (plate 94). These plans were never realised; Robert Adam's revised scheme of 1787 was built instead.

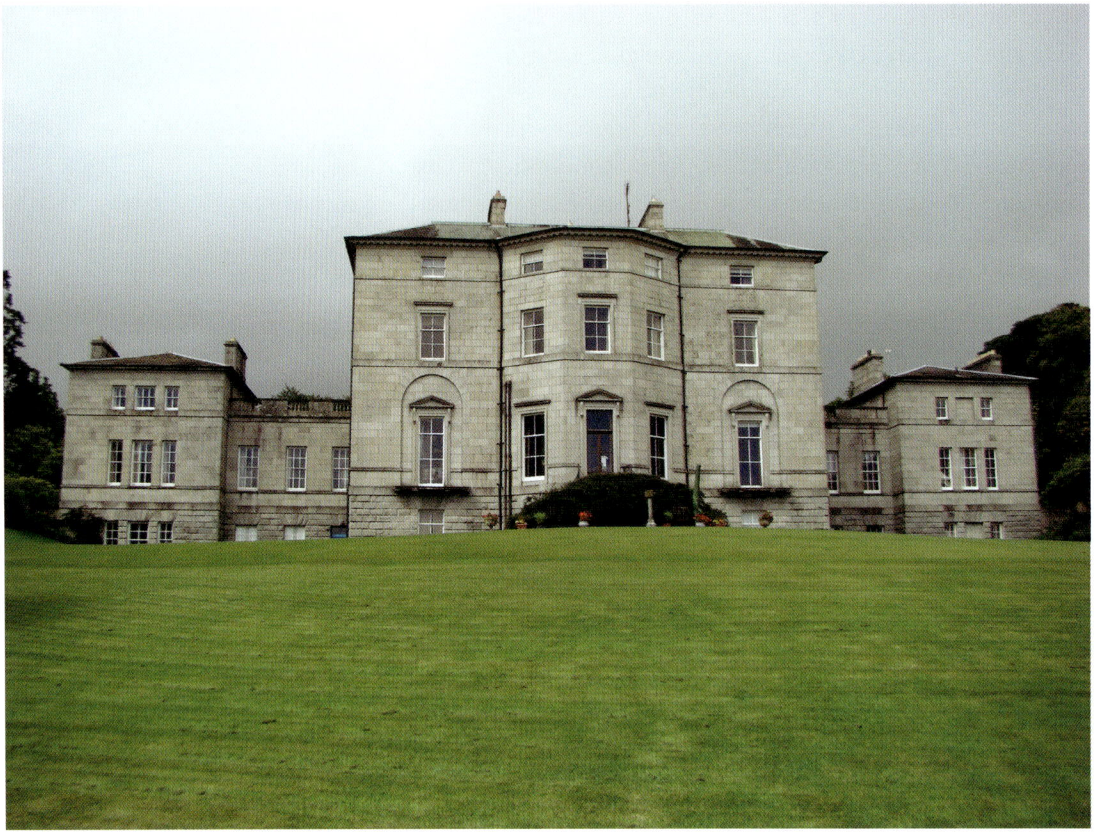

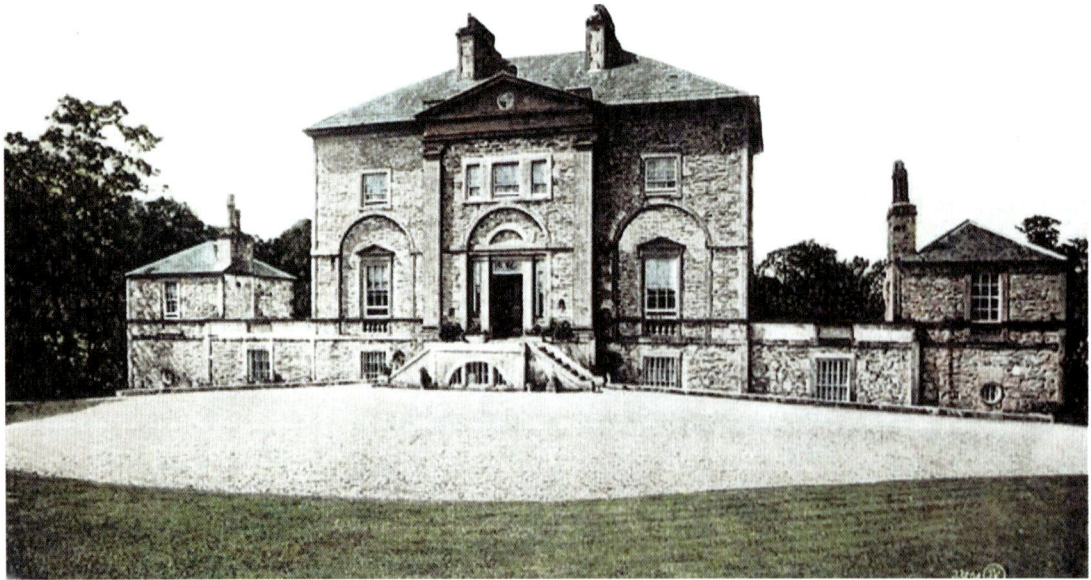

very little storage space was probably not outweighed by its castle-like form, which had become increasingly unfashionable in any case. Although the tower is a potent symbol of feudal power and a romantic icon, in the eighteenth century those with sufficient means preferred to live in a spacious horizontal house with a kitchen, utility rooms, separate quarters for domestic staff and large public rooms with windows that let in plenty of natural light and afforded views across designed parkland.

The alternative to building a completely new house would have been to extend the old tower and incorporate it into a new house, by building on a couple of wings, or surrounding it with a new structure. A number of great houses or castles did exactly that, including Drumlanrig and Culzean. On a more modest scale, several of the Galloway castles described in the Introduction had extensions added over the centuries, including Castle of Park, Lochnaw, Buittle and Kenmure. Local towers Rusco, Buittle and Castle of Park had small eighteenth-century pavilions built as wings to expand the accommodation; small extensions had been part of the plan for Barholm when Ian Lindsay drew up plans for restoration in 1953.

At Barholm Castle the terrain may have militated against extending outwards. Within a few feet of the castle is a steep ravine where rock was quarried for building the tower, and the castle itself is situated on a slope and built on uneven bedrock that surrounds it to the south and west. The logistics of trying to build up and outwards would have been more difficult than starting from scratch, although at some stage simple agricultural buildings had been crudely tacked on to the tower at the north, east and west sides. The situation of Barholm Castle, 400 feet above Wigtown Bay and the Cree estuary, is superb. But the situation of Barholm House at Creetown also afforded lovely views of the estuary (now crowded out by trees) and provided a large area of open, flat ground close to the shore, which Barholm lacked – enough to site a home farm, stables and the usual offices of a country estate, with easy access to the village of Creetown. If the McCullochs had been involved in the smuggling trade, the site in Creetown would also open up the possibility of a small harbour for receiving goods directly from the Isle of Man. As can be seen from the Ordnance Survey map of 1849, there was a jetty directly in front of the house, although it is not known when this was built. The *First Statistical Account* of 1794 described Creetown as having good anchorage, 'to which a ship of 500 tons may come and ride in safety'.

As we know, John McCulloch IV was already planning to entice workers to what became essentially a planned town in Creetown, although there was existing industry in the area: Creetown had at least one mill, a bleachfield

Opposite top.
Kirkdale House as it is today.

Opposite bottom.
A postcard of Barholm House, probably dating to the beginning of the twentieth century.

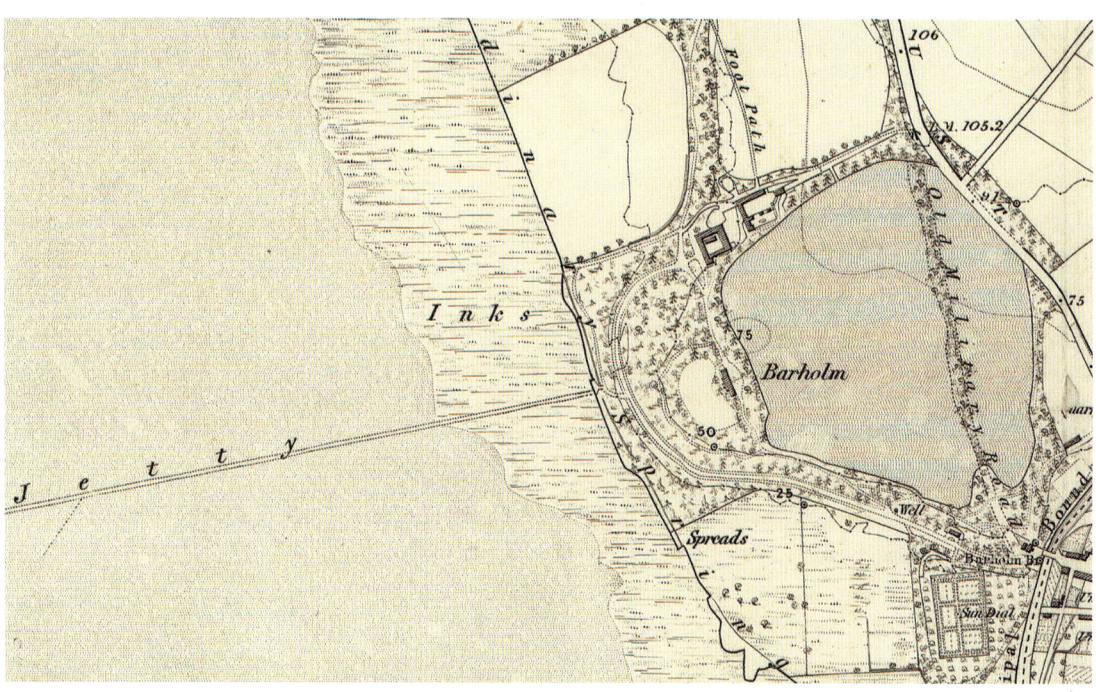

Ordnance Survey map of c. 1849, showing Barholm House, its walled garden, the Mains Farm, a small quarry where stone would have been obtained for the building, and a lengthy jetty stretching out into Wigtown Bay, where ships and smaller craft could have made deliveries and ferried passengers to Wigtown and further afield. If the jetty was in place in the eighteenth century it would have made an ideal point for dropping off smuggled contraband from the Isle of Man.

and a tannery before John McCulloch IV's advertisement of 1771. Both John McCulloch IV and V were 'improving landowners' and, through their influence, more industries were established. The Barony grain mill was built in 1766, a lead shot mill was erected about 1770, in 1771 a tannery was started, and in 1790 a cotton mill, which eventually employed thirty-five women, was built on the Balloch Burn. As a result of all this industry, the population of Creetown rose from 104 in 1764 to 551 in 1793.

John McCulloch VI, the last McCulloch laird of Barholm, was born in 1773. As a child, he resided in Edinburgh with his widowed grandmother, brother and orphaned cousin George during the 1780s. Perhaps he received his education in Edinburgh, where there would have been a far greater choice of schools than locally in Creetown. It is not clear when he succeeded his father. He was a litigious man and attempted – unsuccessfully – to overturn the sale of Barholm lands to Samuel Hannay. In 1791 he wrote a letter (possibly to a fellow landowner in Maxweltown near Dumfries) to complain of 'such an inundation of vagrant Irish families that our Parish is quite distressed with them'. It was only twenty years since the Irish Great Frost, which precipitated a terrible famine and caused over 300,000 deaths in Ireland, a proportionately larger toll than the Great Potato Famine of a hundred years later. Although the economy recovered quite quickly, poverty was rife in Ireland

THE HISTORY OF BARHOLM CASTLE

and many sought a better life across the Irish Sea in southern Scotland. It does not reflect well on John McCulloch, man of the Church, that he sought ways to stop the immigration of Irish workers, describing them as a 'rapidly increasing evil', presumably fearing a drain on the resources of the parish.

It was John McCulloch VI 'of Balhasie' who had Creetown created a Burgh of Barony in 1791, with a charter granted by George III. A ceremony duly took place at the newly erected Barholm House on 13 August 1792, witnessing the Instrument of Sasine. John McCulloch was not being wholly altruistic: the Charter granted certain lands and rights to him and his heirs in perpetuity, as well as giving the inhabitants of Creetown the right to hold courts after electing one bailie and four councillors, and the privilege of holding fairs four times a year, when horses, fish, bread and other marketable goods could be sold. Creetown ceased to exist as a burgh in 1892, under the Police (Scotland) Act.

John McCulloch VI was a proud McCulloch, conscious of his family's status. The execution of Godfrey McCulloch for murder in 1697 had put an

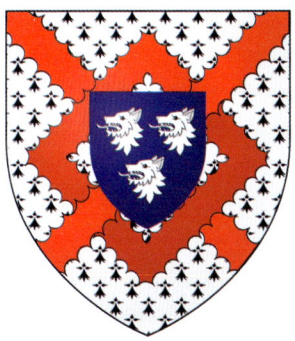

The McCulloch family coat of arms. The motto is *vi et animo* – by strength and courage.

Below.
Robert Adam's plan of the offices and plantation around Barholm House.

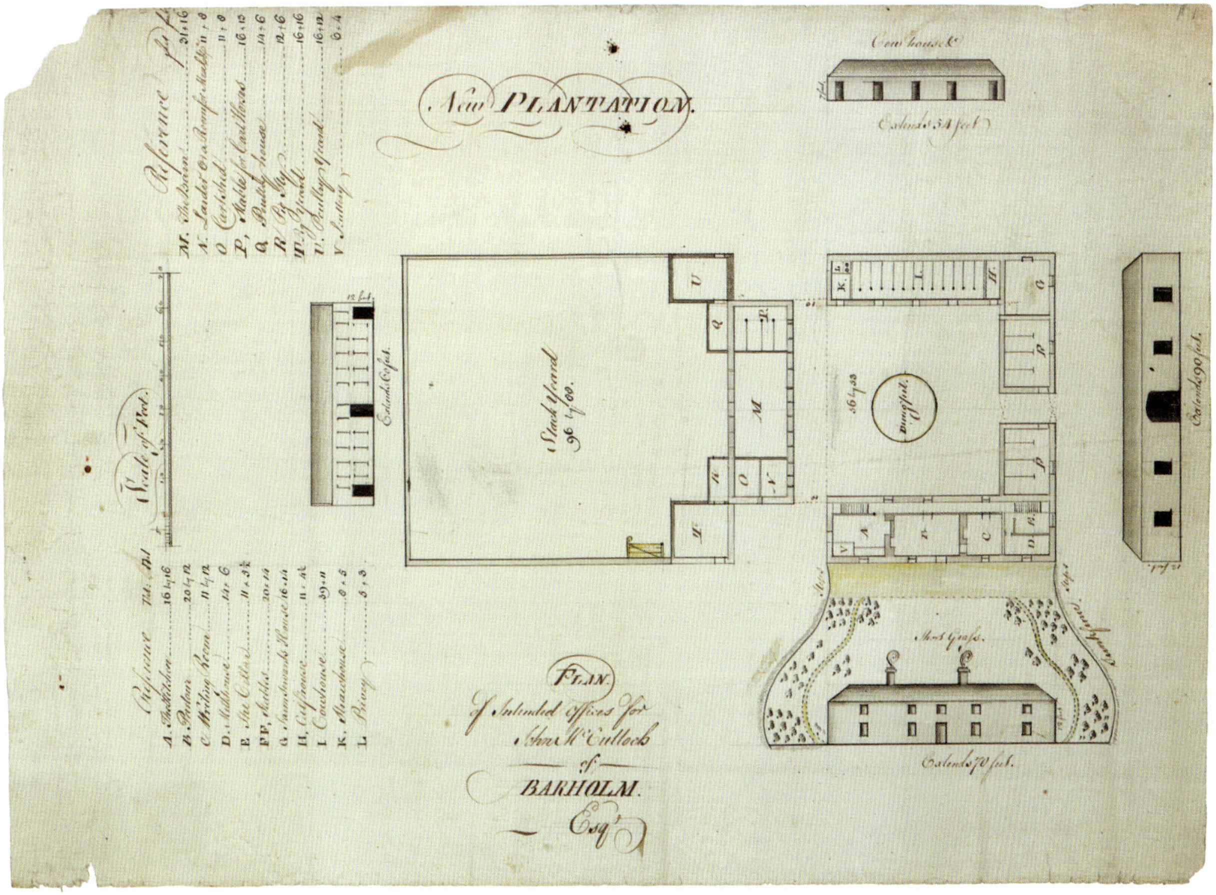

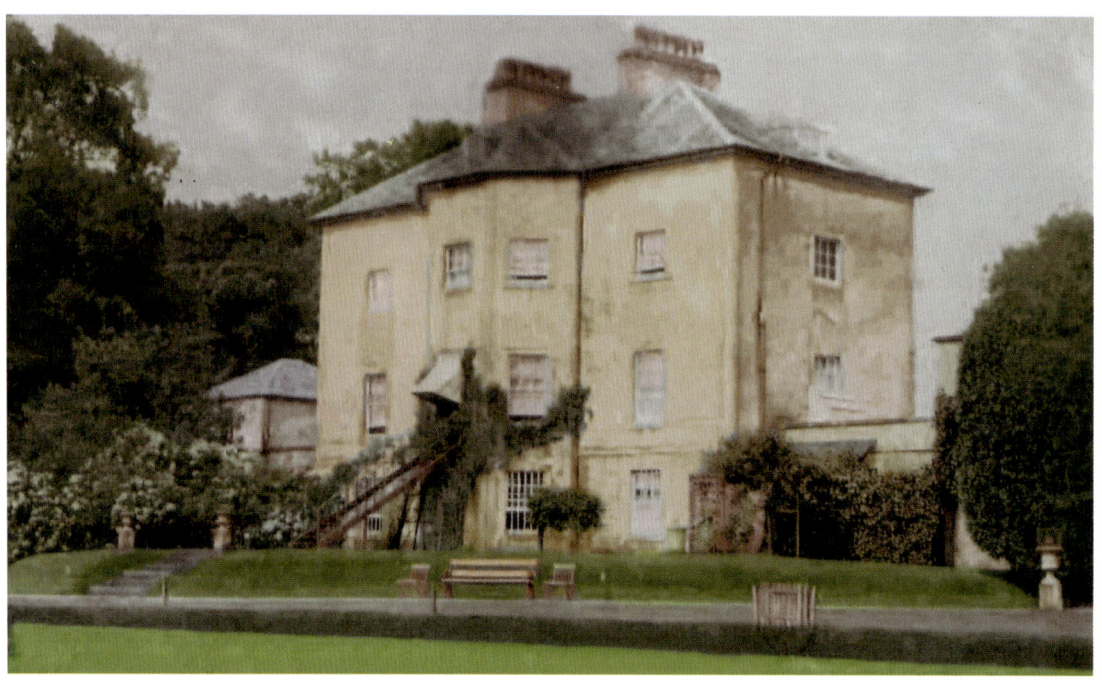

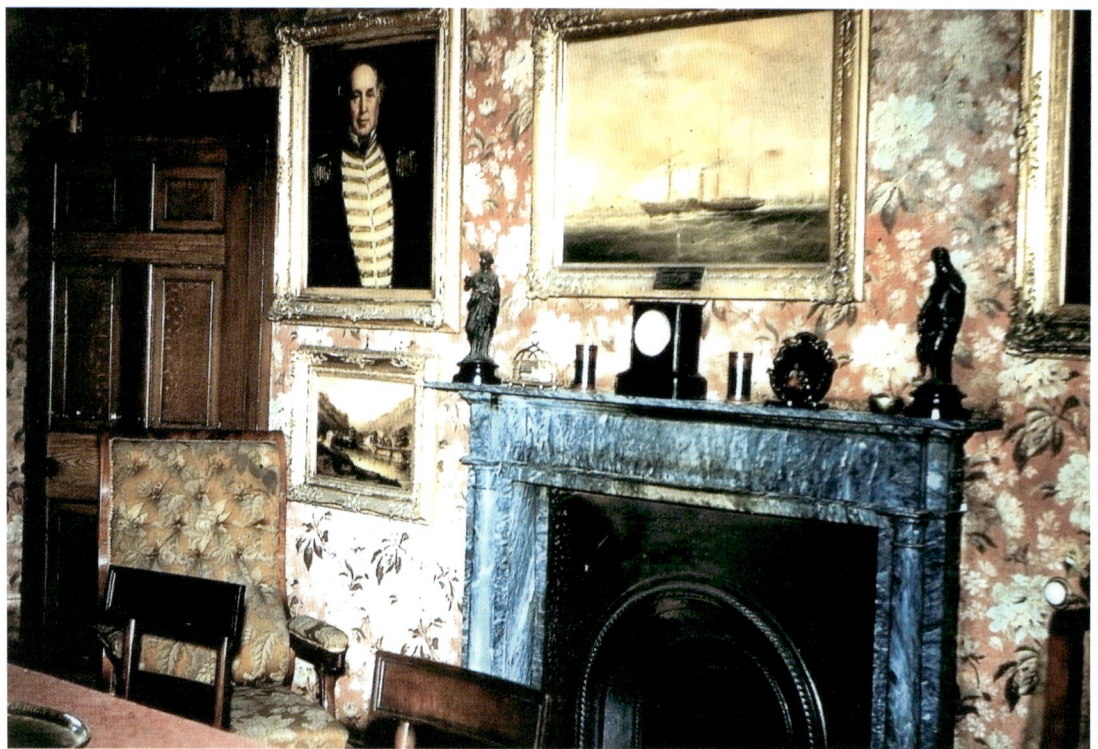

THE HISTORY OF BARHOLM CASTLE

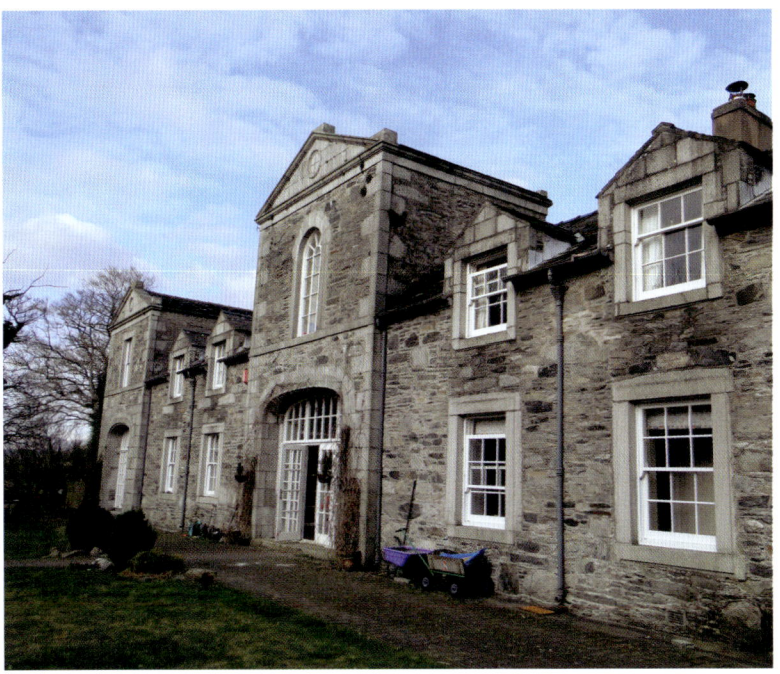

Opposite top.
The rear side of Barholm House. Note the very steep external stair, which appears to be made of metal.

Opposite bottom.
The interior of Barholm House c.1959, showing a portrait of Captain Grant beside the fireplace, with a painting of his ship, the *Berenice*, above the fireplace.

Left.
Barholm House stables, now named Wickham Place and redesigned as a comfortable home; the stables are all that is left of the Barholm estate in Creetown.

end to the title and rank of baronet for the McCullochs, which had been conferred in 1634. On 30 March 1814, John McCulloch VI of Barholm obtained a coat of arms with supporters from the Lord Lyon Office, as the direct lineal descendant of the McCullochs of Muil, a younger branch of the ancient family of McCulloch of Myretoun and Cardoness.

Barholm House was lived in by John McCulloch V, his son John McCulloch VI (1773–1851) and the latter's daughter Isabella McCulloch, with her husband Captain George Grant. After the death of Isabella McCulloch in 1889, Barholm House passed to her two remaining children in succession: first to Jane, who died unmarried in 1899, and then to Agnes, who lived there with her younger cousin Frederick Weekes until her death at the age of eighty-five in 1935. Isabella's sons both pre-deceased her: John, who died unmarried at the age of just twenty-one in 1868, and James, the youngest, who died unmarried in 1888. It seems extraordinary that none of the four Grant McCulloch children ever married. Captain Frederick Weekes inherited Barholm House from his cousin Agnes. He also did not marry, and when he died in 1957 the direct connection to the Barholm McCullochs died with him. When Ambrose Henry Spong, nephew of Freddie Weekes, unexpectedly inherited in 1959 as a young medical student, his mother advised that the only sensible thing to do was to demolish the house, which was riddled with dry rot, and build a new one in its place. He took her advice and that was the end of Barholm House, although its stable block remains, re-purposed in recent years as a comfortable dwelling house.

McCulloch and Barholm Characters

The history of the Barholm McCullochs has so far focused on the direct line of descent. But the younger brothers, cousins and other family members of the Barholm McCullochs, who became seafarers, entrepreneurs, philanthropists and even murderers, also pursued interesting lives, some far away from south-west Scotland. The first character featured in this section was not a McCulloch, but he probably did stay at Barholm Castle, and he became one of the most famous and influential Scots of all time.

John Knox at Barholm Castle

John Knox is reputed to have stayed at Barholm Castle about 1566, after leaving Edinburgh in haste in the wake of the scandalous murder of David Rizzio, the secretary and (possibly) lover of Mary, Queen of Scots. Knox travelled from Ayrshire to Galloway on his way to the Continent and away from Mary. Barholm is thus one of the few Scottish castles in which Knox's arch enemy, Mary, did *not* stay. The evidence, which comes from the *First Statistical Account* of 1794, may seem somewhat circumstantial, but it is a good story, and one that is repeated in every book that mentions Barholm Castle:

> It is a singular fact, which I state on the authority of the present Mr McCulloch of Barholm, that John Knox had his hiding-place in the old tower of Barholm for some time previous to his escape to the continent. This circumstance Mr McCulloch learned from an old man of the name of Andrew Hughan, who was running footman to Mr McCulloch's great great grandfather, and who said that he recollected John Knox's signature on the wall of the small arched apartment or bedroom at the head of the staircase. (*The Statistical Account of Scotland*, 1794)

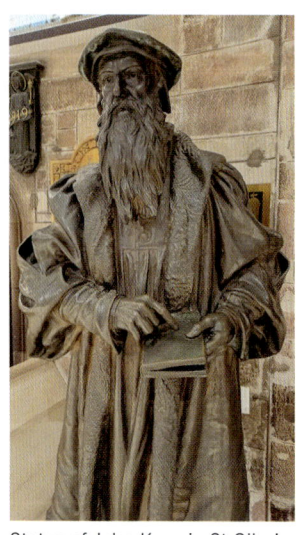

Statue of John Knox in St Giles' Cathedral, Edinburgh.

John Knox's Room is now a cosy little bedroom in the cap house of Barholm Castle, with its vaulted ceiling supporting the floor of the parapet viewing

THE HISTORY OF BARHOLM CASTLE

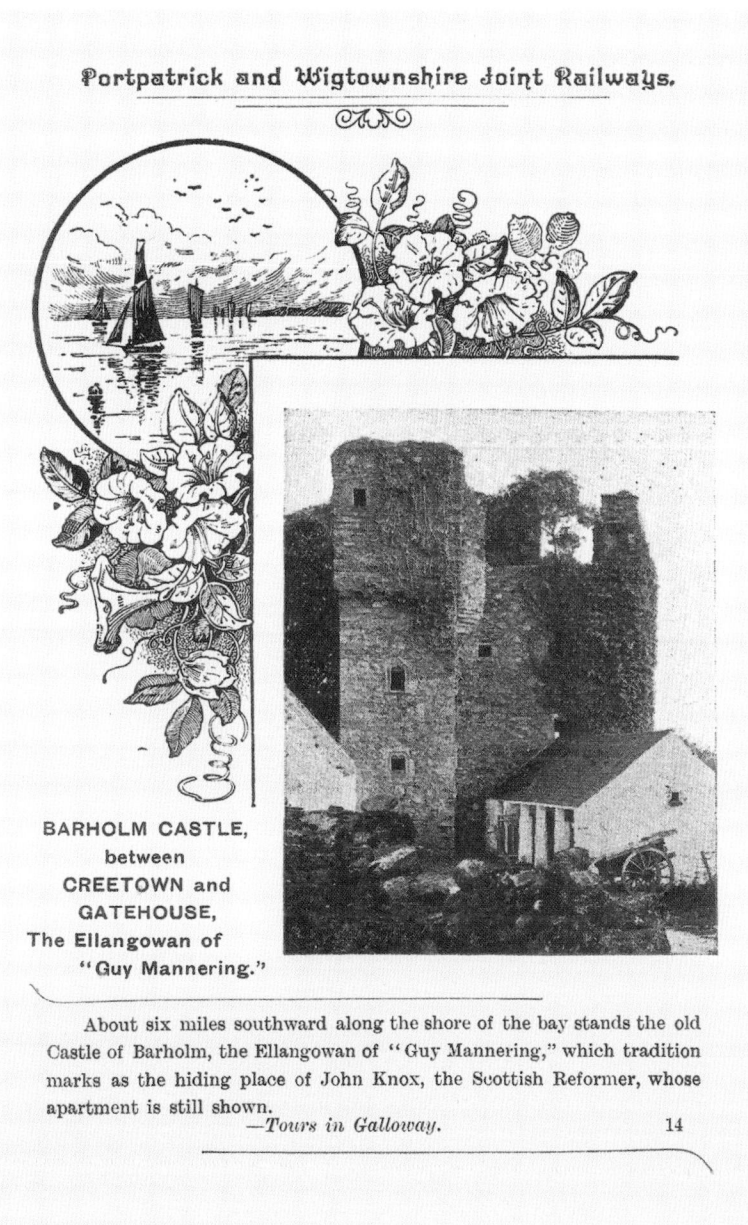

Railway postcard advertising Barholm Castle as 'the hiding place of John Knox'. The postcard dates from around 1904.

platform. The phrase on a railway company advertising postcard, 'the hiding place of John Knox, the Scottish Reformer, whose apartment is still shown', implies that the farmer, or the occupant of the cottage on the east/left of the tower, might have opened up the ruined castle for tourists to climb up the tower, probably on payment of a small fee.

Harry McCulloch, Younger of Barholm

In the middle of the seventeenth century Harry McCulloch, Younger of Barholm, was reported by Morton to be mixed up in an outrage committed on Marion Peebles, widow of Cardoness, the next tower house west of Barholm. (Harry McCulloch does not appear as the 'Younger of Barholm' in Walter McCulloch's (1964) family tree), so this Harry does not seem to have a Barholm 'pedigree'.) This story does not reflect well on the McCulloch perpetrators. The shocking yarn was reported by Robert Scott Fittis in his book *Romantic Narratives from Scottish History and Tradition* (1903). It is reproduced in full here, as the language of Fittis enhances the tale far better than I ever could in a synopsis:

> By the year 1664, the Laird of Cardiness was dead, leaving a widow, Marion Peebles, styled by courtesy of the country Lady Cardiness, and two sons, William and Alexander [Gordon], who lived in family with her at the house of Bussabiel, or Bush o' Bield, in the same parish of Anwoth. This house, which was somewhat of baronial structure, having been probably built for some Laird, and stood in the midst of sheltering trees (hence the word Bush) had been the residence or manse of the famous Samuel Rutherford while minister of Anwoth . . . The Lady Cardiness was now an aged and infirm woman, obliged to walk with a stilt. She was liferentrix of the estate, which, after her death, was to pass to the heir, a young grandson.
>
> The Myretoun baronet [McCulloch] resolved to deal with the strong hand by instituting a 'reign of terror,' which would, he thought, frighten the Gordons out of house and land. With this view, he, on Friday, the 19 August 1664, assembled an armed band comprising his two sons, Godfrey and John, three McCulloch kinsmen, Alexander Ferguson of Kilkerran, and others, and leading them on to Bush o' Bield, began a series of barbarous outrages, which probably could only have been perpetrated in Galloway. The poor old lady was in bed when her enemies came; but this did not prevent them assailing her with blows till she fainted among their hands; and next they pulled down the roof of the room where she lay, with the evident intention of smothering her. Thinking they had effectually disposed of the mother, they fell foul of the son William, 'wounded him dangerously in the arm and hand, to the

hazard of his life, not permitting the servants to give him drink, or go for a chirurgeon to dress his wounds, or administer any kind of help or comfort to him for a long time.' When the gang had done all this mischief, they took their departure. What did the law, so outrageously broken, do in the case? Nothing. William Gordon, dreading a recurrence of the onfall, and justly afraid for his life, thought it best to seek safety in another part of the country, where he remained for some while; but his mother still kept her place.

William was quite right in judging that McCulloch would return like the dog to his vomit. Next year and the year after, he and his emissaries renewed their attacks. On one occasion they treated the old lady in the most unmanly and savage manner; they 'did first beat her almost to death with the stilt wherewith she walked, and then dragged her out of the house and left her upon the dunghill!' This was the form of Galwegian eviction upon impetrated 'Letters'! At another visit, the ruffians behaved with equal inhumanity, dragging the infirm woman out of the house and flinging her down in the open field, and then wantonly breaking and destroying everything within doors. It was perhaps at this time, whilst the house was being ransacked, that Myretoun discovered the title-deeds of Cardiness and took possession of them brevi manu to strengthen his assumed claim. Still, despite all his violence, the lady would not 'flit and remove herself.' So he came back again on another day. She was in bed, and he and his gang 'did keep her from sleep as well as meat; and, further, did throw down water and other liquid matters upon her, so that she was forced to retire and shelter herself within the bounds of the kitchen chimney for her safety.' At intervals of weeks, Myretoun persistently returned, continuing his course of barbarity. He sought to murder the lady's two son's, and seized 'all her rents, corns, goods, and gear, whereupon she could have lived.' In the end, worn out by such prosecution, she burst a blood-vessel and died. Appeal was made to the Privy Council of Scotland, who, after pottering over the case, passed sentence of fine and imprisonment upon the depredators; but it was never carried into effect. Myretoun, however, ceased his attacks, and the Gordons kept possession of Cardiness.

The tale was also reported by Morton (1925–6) in less stirring tones. He brings the story up to date:

In 1668 the McCullochs were back in possession of Cardoness, albeit through violent means. John Gordon was shot and killed by Godfrey McCulloch, son of Alexander McCulloch. This feud with the Gordons had its tragic end in 1690. Sir Godfrey M'Culloch went to Bussabiel and asked to see William Gordon, and when Gordon appeared McCulloch shot him in the thigh and he died in a few hours. McCulloch fled abroad to escape justice, but returned in a few years. One Sunday worshippers in a church at Edinburgh were startled by the cry, 'Steek the door! There's a murderer in the kirk.' It was from a Galloway man who recognised McCulloch. The latter was brought to trial and executed on 25th March 1697.

Bonnie Bess: Elizabeth McCulloch of Barholm and the Walter Scott connection

In 2021 James Shirreff came to visit us at Barholm, bringing with him the diary of his great grandmother Elizabeth Davidson Shirreff, which shed interesting light on Barholm connections with Sir Walter Scott. James's fourth great-grandmother was Elizabeth McCulloch (Bonnie Bess) of Barholm who married Andrew Livingstone of the Airds in 1805. Elizabeth McCulloch of Barholm was a noted beauty, the toast of four counties. She died young, leaving two daughters: Frances, who married Nathan Milligan, and Mary, who died unmarried. Elizabeth Shirreff's diary noted:

> Sir Walter Scott's brother married a McCulloch, cousin or aunt to 'Bonnie Bess'. Sir Walter was intimate with the Barholm family and his novel *Guy Mannering* is supposed to [illegible] its scene near there. He mentions having got a ballad from Mr Andrew Livingstone. Barholm was sold recently to the Duke of Bedford. Elizabeth McCulloch heiress of Barholm married David, second son of McCulloch of Ardwell. Their son John married Mrs Cullen of Argrennan. Their son was John. His son was John.

As Elizabeth correctly recorded, Captain Thomas Scott (1771–1823), brother of Sir Walter Scott, married Elizabeth McCulloch (1776–1848) in Dumfries on 15 December 1799, before migrating to Canada. It is also interesting to note that two branches of the McCulloch family – Barholm and Ardwell – were joined in marriage, although this is not represented in Walter McCul-

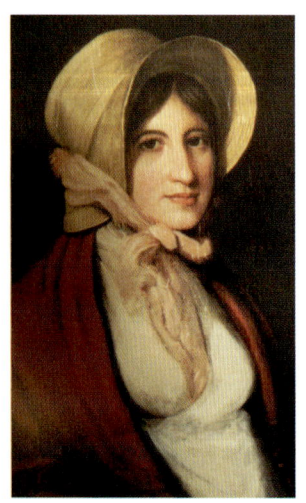

'Bonnie Bess', Elizabeth McCulloch of Barholm (1787–1809); she was said to be the toast of four counties.

loch's (1964) family tree of the Barholm McCullochs.

Elizabeth Shirreff claims in her diary that the Duke of Bedford bought Barholm. James Shirreff assumed that his great-grandmother was referring to Barholm Castle and discounted the story as a myth or muddle of some kind. When he asked us about it, we agreed that there was no evidence of any sort to back up the claim. However, when we investigated the history of Barholm House and the two sisters Agnes and Jane who lived there, we discovered quite recently that the Duke and Duchess of Bedford were regular visitors to the area, and that the Duke bought part of the Barholm House estate in about 1930. So Elizabeth had been correct after all.

George Perrott McCulloch, USA entrepreneur and slave owner

On 7 November 1856, George Perrott McCulloch (1775–1858) wrote a letter from his home in Morristown, New Jersey, USA, to 'McCulloch of Barholm'. The purpose of his letter was to introduce his son-in-law Mr Miller, 'now travelling in Europe to restore a health impaired by 12 years of toil at Washington as Senator of the United States'. Mr Miller's daughter, Miss Miller (the granddaughter of George P. McCulloch) was travelling with her father and, he said, 'feels a natural curiosity to view the localities and learn the traditions of her ancestry'. George McCulloch began his letter:

> Sir,
> It must be about 73 years since your father John, your uncle James and myself (the son of your granduncle William who died in India) resided in Edinburgh, with our venerable grandmother, widow of your great-grandfather, John McCulloch of Barholm. We became separated; your father returning to Galloway, your uncle entering the Naval Service and I emigrating to England.

It seems from the salutation that George believed that his cousin John had had a son – but John McCulloch VI had six daughters and no (legitimate) sons. His oldest daughter, Isabella, inherited Barholm House and the estate, so she would have received the letter, if it reached Barholm. It is not known whether Miss Miller and her father were able to visit Isabella and her husband, Captain Grant.

George Perrott McCulloch was born in 1775 in Bombay (Mumbai), where his father William was a major in the East India Company. Orphaned at the age of nine, he was sent to Edinburgh to be raised by his widowed paternal grandmother, formerly Miss Elizabeth Cutlar of Argrennan. In the Post Office Directory for Edinburgh of 1784–5, Mrs McCulloch of Barholm is listed as living in Crichton's Street, which is near George Square. George lived with his grandmother and McCulloch cousins James and John while he attended Edinburgh University and became fluent in French, German, Spanish and Italian. After university he went south to London and there joined the successful business of Johnson, MacCulloch and Law. Francis Law was the brother of Napolean's aide-de-camp, Count de Lauriston. George travelled extensively in France and Spain during the early years of the Napoleonic wars, often disguising his nationality. In 1800 he married fifteen-year-old Louisa Edwina Saunderson. Six years later they left England with their two young children, Francis Law and Mary Louisa, to settle in America.

The McCullochs probably emigrated to America because of the difficulties of carrying on trade during the wars between Britain and France. George arrived in New York with substantial means and chose property in Morris County probably because French emigrants had already settled there. In 1810 George purchased 25 acres in the rural community of Morristown and there he and Louisa had an imposing Federal-style brick house built – Macculloch Hall – which is now run by a charitable trust and open to the public as a historical museum. The McCullochs expanded their mansion in 1812 and 1819, tripling its size, as their family's prominence in local, state and national affairs grew. George and Louisa brought with them to Macculloch Hall three enslaved adults, Cato, Susan and Betty, and a toddler, Emma. It is not known when or where the slaves were purchased, but George would 'loan out' Cato to do work for other people in Morristown and charge them for Cato's services. From the family Bible there is evidence that three children were born into slavery in Macculloch Hall in the early 1800s. When one of these children ran away in 1832, George recorded aggrievedly, 'Our own boy Henry, after receiving our ample provision of winter clothes, a set of new shirts, hat and coat, went off 3 weeks ago and is supposed to be now in New York. Peace and prosperity be with him – we are well rid of him and did not move a single step to recover him.' George must have been aware, both from his time at Edinburgh University and in London, and his elevated position in New Jersey society, that the concept of slavery was increasingly frowned upon on moral and humanitarian grounds, but clearly he felt justified in owning his own.

George Perrott McCulloch (who changed the spelling of his name to MacCulloch during his time in the USA) had been sponsored in his ventures in the USA by Napolean Bonaparte. Red flannel underwear was often worn

THE HISTORY OF BARHOLM CASTLE

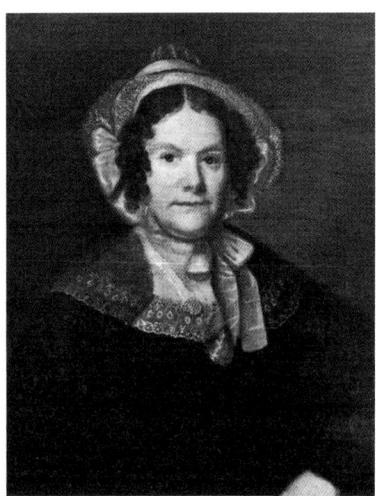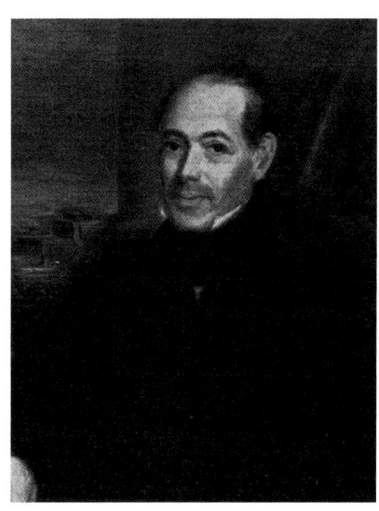

George Perrott McCulloch and his wife Louisa, c. 1850.

in New Jersey to resist the severe winters. When George learned that Napolean was bound for Russia, he determined to send him an outfit of red flannel drawers. He sent it through his firm of Johnson, MacCulloch and Law, so that during the freezing conditions of the retreat from Moscow, Napolean could wear warm underwear for some comfort. John Brett Langstaff, who wrote *New Jersey Generations: Macculloch Hall, Morristown* suggests that it was George's Scottish character that shaped him in later years:

> The rugged forebears of George Perrott MacCulloch are an essential consideration in a full understanding of the character of the man who built Macculloch Hall. The rough nature of the Scottish homeland and the severity of the elements with which the inhabitants had to contend fitted Scotsmen for exploring the wilds of North America and often made them daring adventurers in the realms of finance. Courage to face the seemingly insurmountable obstacles of nature, a readiness to fight for what they possessed and an intellect sharpened by friction between the clans were all part of MacCulloch's character as a Scotsman. (Langstaff 1964, p. 20)

Captain William McCulloch, aka 'Flogging Joey'

There is a wall tablet on the outside of St Leonard's Church in Deal, near Canterbury, Kent, which reads simply: 'Sacred to the memory of Captain William McCulloch RN of Barholm, Scotland, who departed this life on the

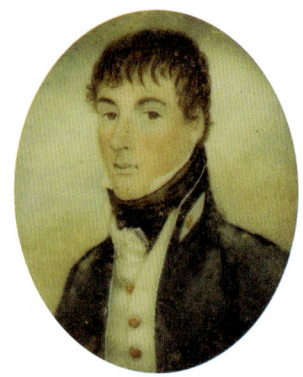

Captain William McCulloch, 'Flogging Joey'.

25th October, 1825. Age 45 years.' William McCulloch was the younger half-brother of John McCulloch VI. His father, John McCulloch V, had married twice, the second marriage being to a servant girl, Jessie MacFarlane, who bore seven children, including one with the interesting name of Samuel Hannay McCulloch. Captain William McCulloch was in command of the Coastal Blockade in Kent from 1823 to 1825, instituted to put an end to the practice of smuggling, and was Captain of the *Ramillies* at the time of his death. He married Jane Osborne in the Island of Antigua; she bore eight children and outlived him, dying in 1853. There is a small book about Captain McCulloch, called *Flogging Joey's Warriors*, in which the author raises a question about McCulloch's character, 'Often depicted as a flogging enthusiast of the old school, sometimes described as a sadistic monster, was he, perhaps more credibly, simply a hard man doing a difficult task to the best of his undoubted ability?' (Douch 1985, p. 60). Whatever the truth, it is telling that his nickname was Flogging Joey, and after his death, when it was usually the custom that fellow officers would contribute to a memorial for their captain, nothing was forthcoming. Douch also questions the probity of the Barholm McCullochs back in Scotland: 'It is a curious paradox that although Captain McCulloch became the scourge of the Channel smugglers, both his father and grandfather had no doubt, like all the Galloway lairds, made a very good thing out of smuggling on the Solway which was particularly well adapted for the purpose, being very near the Isle of Man where bulk cargoes could be broken up and the goods transferred to smaller, faster craft' (Douch 1985, p. 64).

James McCulloch, illegitimate son of John McCulloch

James Murray McCulloch was born out of wedlock to John McCulloch VI in 1804, at Barholm House. This was ten years before John's marriage to Agnes Stewart; the name of James's mother is not recorded. James studied medicine at Edinburgh, where, despite his illegitimacy, he stayed in the McCulloch family home in Albany Street. He was attached to an army medical unit in India before returning to Scotland and setting up a medical practice in Dumfries, where there was a terrible outbreak of cholera. 'Undaunted by the virulence of the disease, Dr McCulloch laboured heroically among the afflicted, visiting as many as seventy patients in a single morning. Indeed, so fatigued did he become that he often had to be carried bodily up a flight of stairs in order to provide medical treatment and comfort to his patients.

Although he himself contracted the disease, he was fortunate in recovering' (Macleod 2024, p. 18).

Such was his reputation that, in 1859, the ladies of Dumfries subscribed for a handsome carriage and pair for the doctor. The accompanying fulsome address expressed admiration for his 'uniform manliness and nobility of bearing to all who came under his care. To belong to the suffering is the only qualification to secure your interest. The life of a poor man is as valuable in your estimation as the life of a noble.' Ironically, had James not been born illegitimately, as the first-born son of John McCulloch he would have become laird and inherited Barholm Estate. Perhaps this awareness of missing out on his rightful position in life gave him a compassion for others less fortunate.

In 1834, James married Mary Ellison Lafone in Aigburth, Lancashire. Mary, like her husband, became a social reformer, and in 1867 her name headed the first petition on women's suffrage to be presented to the UK Parliament. James spoke regularly at Temperance Society meetings and was its vice president. When he died in 1888, aged 84, his death certificate gave his father's name as John McCulloch, landed proprietor.

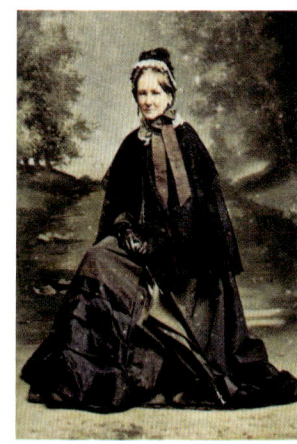

Mary Ellison Lafone, women's suffrage campaigner and wife of Dr James McCulloch.

Captain Grant of the Indian Navy

George Grant was born in 1793. He married into the McCulloch family when he wed Isabella McCulloch, oldest daughter of John McCulloch VI, in 1845. He and Isabella were living at Castlewigg, Wigtownshire, at the time of the 1851 census. This was the year that Isabella's father died. It was reported that Captain Grant, on the death of John McCulloch VI, purchased Barholm, but one assumes that the house passed directly to Isabella on the death of her father. Perhaps it was mortgaged and the new son-in-law paid off the debt.

McKerlie gives an account of Grant's life:

> George Grant entered the Indian Navy in 1810. In 1813, he was promoted to the rank of lieutenant. He distinguished himself in the capture of piratical vessels, and received the thanks of the Indian Government, also for his services against Kundorna Fort in Kattywar. He was presented with a valuable sword, etc., by His Highness the Guikwar, and the rank of Captain in his service, conferred. He was assistant in the Master Attendants' department, and afterwards Acting Master Attendant at Bombay. In 1837 he

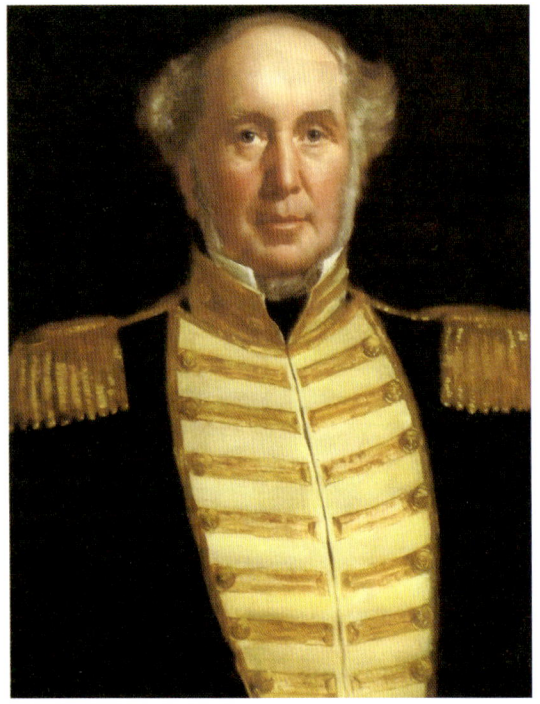 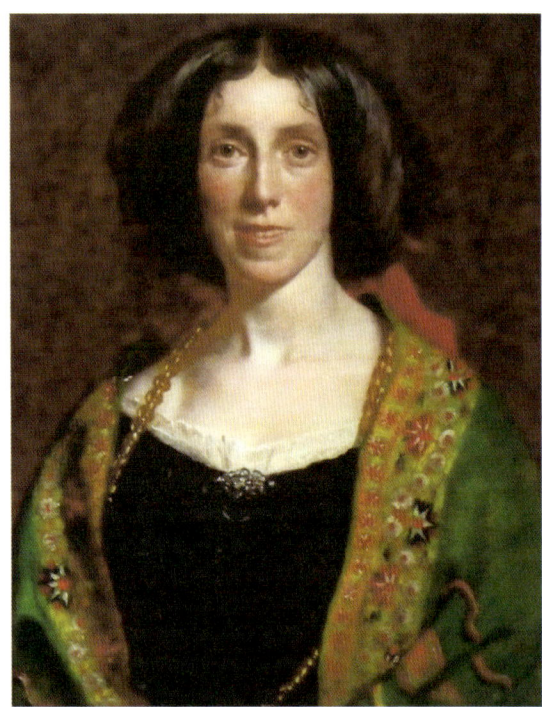

Left.
Captain Grant, painted in 1858 by William Crabb (1811–1876).

Right.
Isabella McCulloch, wife of Captain Grant.

commanded the steam frigate Berenice, 'built on the Clyde for the Indian Navy' and took her out. She was one of the two vessels which first steamed round the Cape of Good Hope for India. At this time, when a boy, we became acquainted with Captain Grant, that is before he left Culderry, near Garlieston, which he rented, to proceed to India in the 'Berenice'. Captain Grant died at Barholm on the 22nd September 1874, aged 82, leaving Isabella a widow for the next fifteen years. Their oldest child, James, predeceased her by one year. (McKerlie 1878, p. 255)

Jane and Agnes Stewart McCulloch Grant

Sisters Jane and Agnes McCulloch Grant were the daughters of Isabella McCulloch and Captain Grant. Jane inherited Barholm House in 1889, and on her death ten years later, the inheritance passed to Agnes. The sisters were conscious of their family history and together they organised a memorial stone to their forefathers, and in particular their grandfather, John McCulloch VI, which they placed in the ruined old church of Kirkdale. It reads:

THE HISTORY OF BARHOLM CASTLE

> Within
> This old Church of Kirkdale
> Has been for more than 500 years
> The family burying place
> Of the McCulloch of Barholm
> Formerly of Muille in Wigtownshire
> The last who was laid to rest here
> Was John McCulloch of Barholm
> On October 4th 1851
> Who died at Barholm House on Sept 30th
> Aged 78 years
> This stone is erected to the memory
> Of their forefathers
> By his granddaughters
> Jane H. McCulloch Grant
> And Agnes Stewart Grant
> 1897
> One generation passeth away
> And another generation cometh
> But the earth abideth for ever
> ECCL 1st Chapter Verse IV

The Rev. John Muir, in the *Second Statistical Account* of 1845, describes the place in poetic terms:

> It is impossible to conceive a more lonely spot. The situation is solitude itself – remote – romantic – placed in a dreary vale, in the bosom of a wood surrounded by hills – in sight of the troubled ocean, and within hearing of the ceaseless wailings of the stream. Nothing can surpass the solemnity of the scene. In entering the churchyard, the living feel as already in communion with the dead, and behold, in the most striking manner, 'How still and peaceful is the grave.'

By the time Agnes owned Barholm House in 1899, funds were becoming low. Luckily, she had friends in the Duke and Duchess of Bedford, who for several years rented nearby Cairnsmore House for shooting holidays. The Duchess, who took up flying at the age of 63, used to fly up to Galloway from Woburn Abbey, and was known as the 'flying duchess'. By 1930, the Duke of

Miss Jane McCulloch Grant.

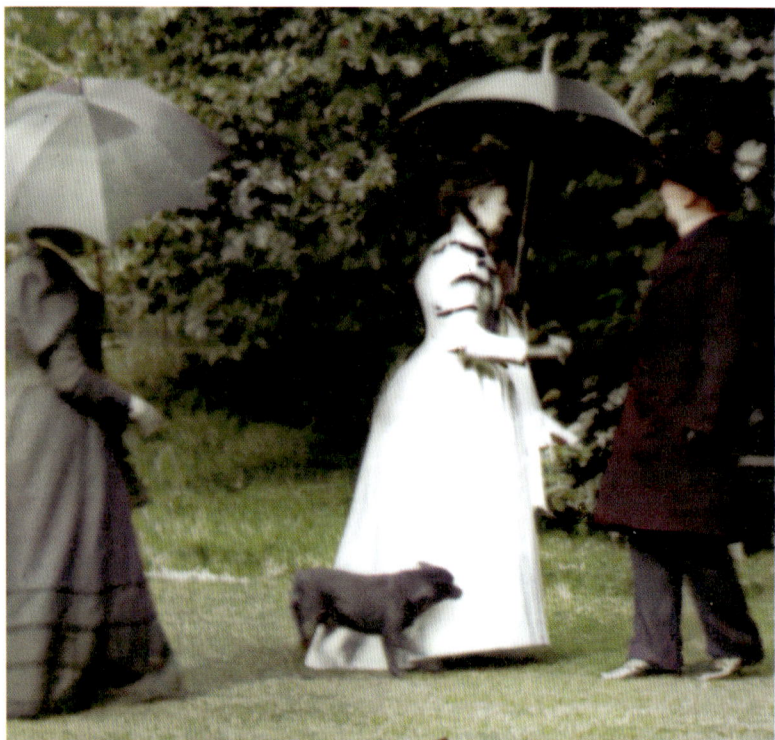

Bedford had bought most of the estates of Barholm. The money from the Duke was probably only a temporary fix, however. Agnes was left with much reduced land. She now owned only Barholm House and its home farm, along with some land and a manure depot on Chain Road, and a part of the Ferry Croft. A large house such as Barholm can be easily maintained only if supported by regular income from land rent and farming. Barholm Farm was probably not sufficient on its own. By the time Freddie Weekes inherited Barholm House in 1935 the rot had, literally, begun to set in.

Frederick Wickham Weekes: the last of the line

Frederick Wickham Weekes, born in 1869, was the cousin of John, Jane, Agnes and James Grant, who were the children of Isabella McCulloch and Captain George Grant. Frederick's mother was Elizabeth McCulloch, sister of Isabella, and his father was Captain Weekes of the 78th Hussars.

In 1891, according to the census, Frederick was living in Cathcart in Renfrewshire in lodgings. His occupation was given as civil engineer. His

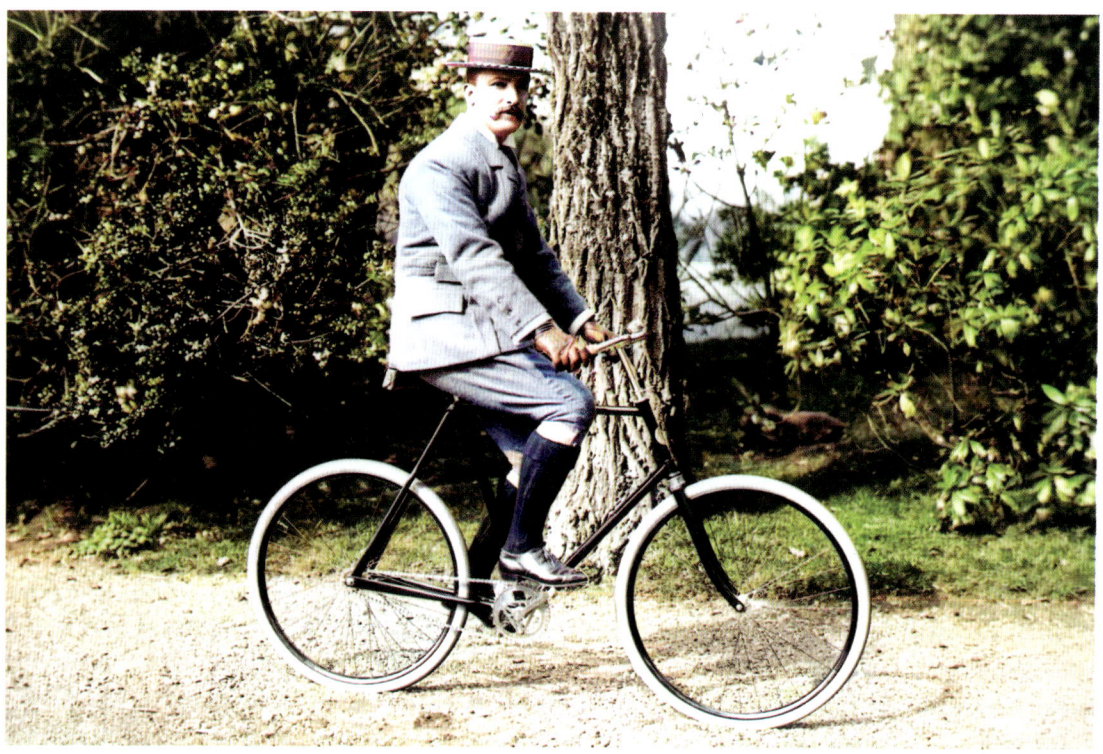

mother, Elizabeth McCulloch, was living ('on private means') at that time in Barholm House, where her niece, Jane, was head of household. Her sister Agnes was not living there at this point; she succeeded to the estate in 1899, on the death of Jane. By 1901 Frederick was resident in Barholm House, living with his cousin 'on own means'. The private means was not an inheritance from his widowed mother, who did not die until 1906, but perhaps she gave him an allowance. In 1914 the First World War broke out. Frederick was probably too old by then, at 45 years old, to answer the first call-up, but a Frederick Weekes served in the Black Watch 1917–19. He was Chairman of Dumfries District Board of Control 1925–30 and thereafter Convener of the Stewartry. In 1934 Frederick was appointed deputy Lord Lieutenant of the Stewartry of Kirkcudbright. Weekes' lasting legacy for Creetown was his donation in 1936 of land at 'The Hollow', to be turned into playing fields to commemorate King George V. These were part of a commemorative scheme involving 471 playing fields across the United Kingdom; the playing fields are still in use by the community and have been changed and upgraded several times since they were opened.

When Frederick inherited Barholm House from his cousin Agnes Stewart

Frederick Wickham Weekes wearing a boater and knickerbockers, posing on a pedal bicycle with pneumatic tyres; these details date the photograph to the late nineteenth century or early twentieth century.

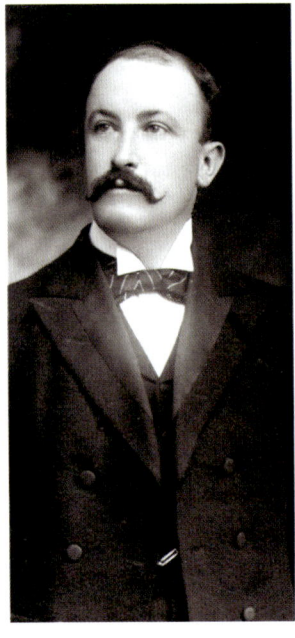
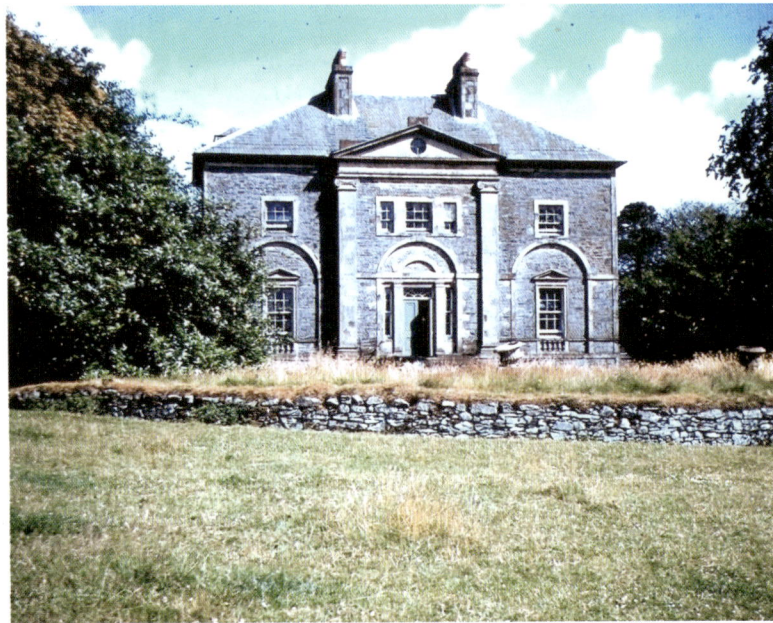

Above.
Frederick Wickham Weekes.

Right.
Barholm House in 1959, minus its two wings, shortly before it was demolished.

McCulloch Grant in 1935, he was the last child of a Barholm McCulloch parent to live there, and indeed the last inhabitant of the house. The name McCulloch died out with him when he died, unmarried, in 1957 at the age of 88. He was described on his death certificate as a Captain in the Seaforth Highlanders. Barholm House was by then in a bad state and had had its wings removed at some point. It was demolished in 1959 by A. Henry Spong, who inherited it as a young medical student and was not able to afford to repair the building, which by then was riddled with dry rot. Dr Spong and his wife Gill now live in Plymouth, but visit Galloway regularly.

Barholm Castle in the Nineteenth and Twentieth Centuries

Malcolm McLachlan Harper, in his *Rambles in Galloway*, noted that 'The castle is in pretty good repair, but is uninhabited and appears to be used as a lumber store by the farmer of Barholm, whose dwelling-house is contiguous' (1876, p. 112). Once a building is abandoned, it does not take long for degradation to kick in. Already by 1782, as the Ainslie map describes, Barholm Castle was in ruins; it would remain so for more than another 200 years.

In 1878, McKerlie wrote about Kirkdale House, owned by the Hannay family:

THE HISTORY OF BARHOLM CASTLE

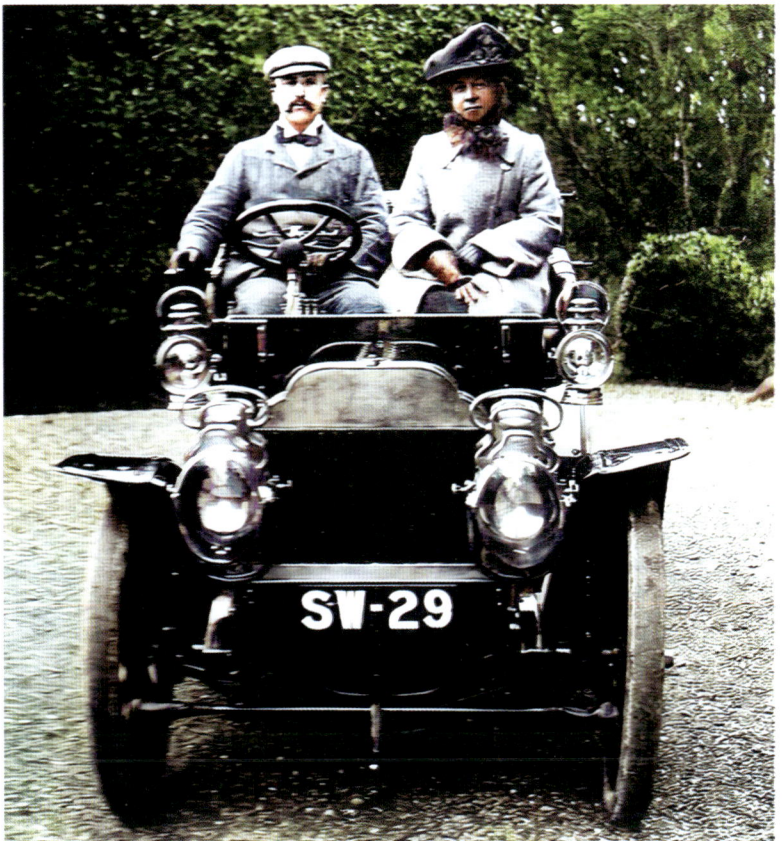

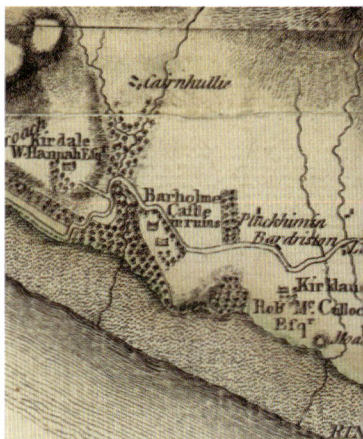

Left.
Freddie and Agnes in a motor car, probably around 1900.

Below.
Extract from John Ainslie's *Map of the County of Wigton*, 1782, which labels Barholm Castle as 'in ruins'.

The old castle of Barholm belongs to this property, the name being absorbed in that of Kirkdale, although nominally transferred to the more modern residence of the McCullochs, now Grant, at Creetown. The castle is east of Kirkdale House, and is now merely an outward shell, excepting a staircase, the interior fittings having disappeared. In it John Knox was hiding until he escaped abroad. There can be no doubt of this, considering the source from which this information is derived, and to those who value the vast services of that great man, the interest in the old building should be greatly enhanced. [McKerlie seems to imply Sir Walter Scott here.] The castle (as well as the one at Carsluith) is supposed to be the 'Ellangowan' in Sir Walter Scott's 'Guy Mannering', and having belonged to the McCullochs, it seems to us the place. The situation and scenery around strengthens this opinion.
(McKerlie 1878, pp. 256–7)

Right.
Nineteenth- and twentieth-century graffiti was inscribed in the turret stair walls but has now been plastered over.

Below.
The intrepid climbers standing on the top of the cap house may have scratched their names on the walls of the turret staircase. The barn to the right of the picture had gone by the 1930s, as can be seen in the next photograph, although the 'standing stones' which held up the roof remain.

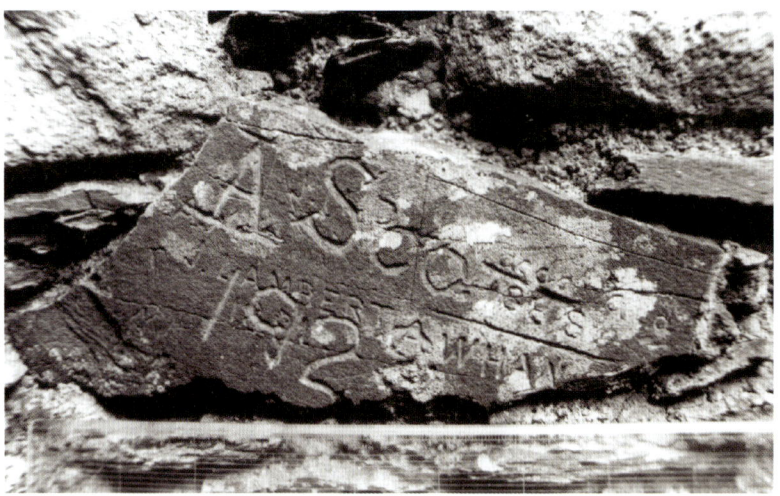

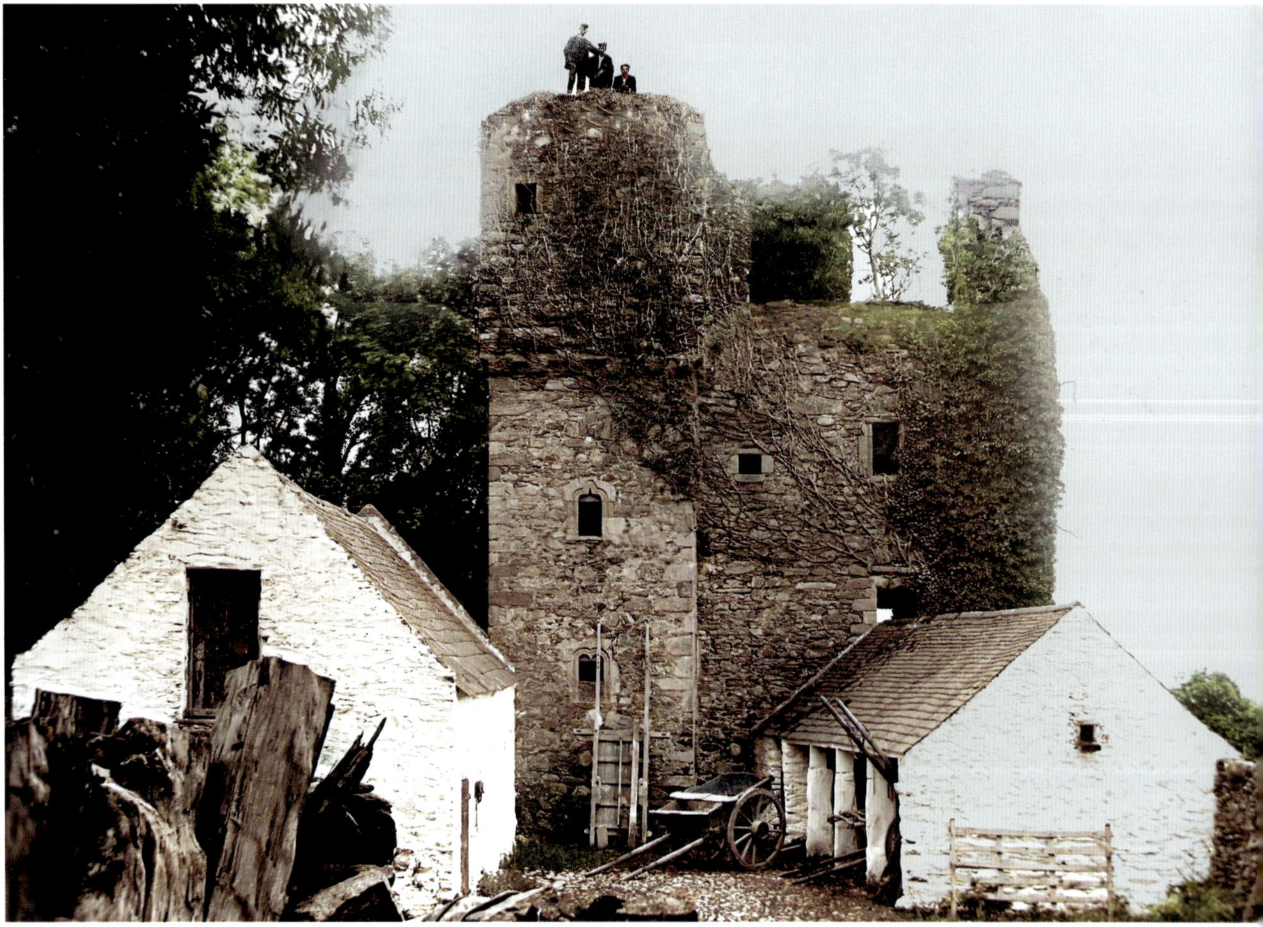

THE HISTORY OF BARHOLM CASTLE

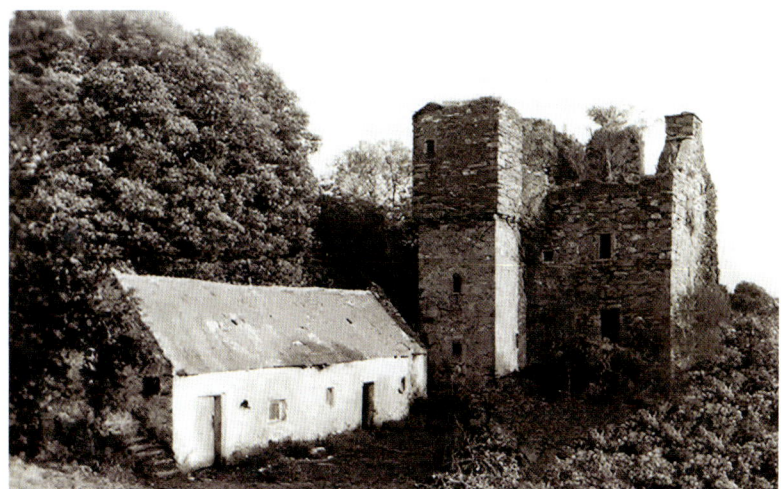

Barholm Castle in the 1930s. The building which formerly abutted the castle (under the big window) had been demolished by this time.

Barholm was one of many romantic ruins that were explored by local children and antiquarians alike. In 1925, A.S. Morton led a field study trip to Barholm Tower for the Dumfries and Galloway Natural History and Antiquarian Society and described the details of the trip, along with a history of Barholm, in the society's *Transactions* (Morton 1925–6, p. 232). The trip included a visit to Cairn Holy chambered cairns nearby and was no doubt popular. In the report of his vote of thanks, the Rev. Dr J. King Hewison was of the opinion that 'Galloway did not fully appreciate what had been done by the McCullochs, McLellands and Gordons in years gone by. They might not have been there that day had it not been for the bravery of those families, for they did much to bring peace to the country.' The Minister was referring to the Covenanting tradition, which brought a degree of religious freedom (if not tolerance) to Scotland.

On 13 March 1937, in recognition of its significance, the Commissioners of His Majesty's Works and Public Buildings sent notice to Colonel Frederick Rainsford Hannay, then owner of Barholm Castle, that they intended to include Barholm Castle on the list of scheduled monuments of national importance. It remained on the schedule until well after the restoration in 2008.

Barholm Hamlet and the 'new house' down the track

In 1959, Major Ramsay Rainsford Hannay, the next owner of Barholm Castle and Farm, decided to build a new house for his factor, Barholm Farmhouse having apparently become unfit for habitation. The buildings abutting Barholm Castle were demolished and a track made leading south past the castle, to a flat area of ground where a small house was constructed (this replacement for the old farmhouse was confusingly called Barholm House but it is not related to the mansion of the same name in Creetown). Not long after the new Barholm House was built, however, Rainsford Hannay sold it, along with Barholm Castle and Barholm Farm, to Billy Hogg and his wife Ann Hogg, née Younger. They added an extension to the house, doubling its size to accommodate their young family. Mrs Hogg was a member of the Younger family of Benmore, the Argyll estate whose magnificent botanic gardens were gifted to the nation in 1929 and are now open to the public as an outpost of the Royal Botanic Garden Edinburgh. The large mature rhododendrons at the shallow end of the ravine came from Benmore. Ken Cox, the rhododendron expert from Glendoick Gardens, came to Barholm a few years ago and identified our huge rhododendrons as Benmore hybrids.

In the ravine, we often dig up glassware which was probably thrown there by the inhabitants of the cottage that was demolished to make way for the track down to the new Barholm House. Judging by the glassware that we find, we imagine that they drank gin and ate lots of meat-paste sandwiches.

The 1950s was a time of utter contempt for old buildings. After the Second World War, people wanted to live in convenient new houses, not ancient stone cottages. There was wholesale destruction of Scottish – and British and European – heritage buildings throughout the 1950s and 1960s. Fortunately, the original Barholm Farmhouse was not demolished in the 1950s when the new Barholm House was built. In the 1960s, Billy and Ann Hogg undertook a restoration of the almost derelict farmhouse. Later, their daughter Janet and her husband Will Barrie moved in; they have been farming the land around Barholm ever since, although in 2015 they moved out of the farmhouse into nearby Barholm Byre, which they restored from dereliction into a comfortable home. Derelict Barholm Mill was also made into a comfortable home by Michael and Sarah Crothers not long after we restored Barholm Castle.

When we first saw the ruin of Barholm Castle, there were only three inhabited houses along our track: Barholm Cottage, Barholm Farmhouse and Barholm House. Now, with the addition of Barholm Byre, Barholm Mill and Barholm Castle, all restored from ruins this century, there are six: a little hamlet. In the eighteenth and nineteenth centuries there would have been a series of mills along the valley, many providing flour for bread. In 2005 Dr David Hannay restored nearby Kirkdale sawmill, which had been servicing

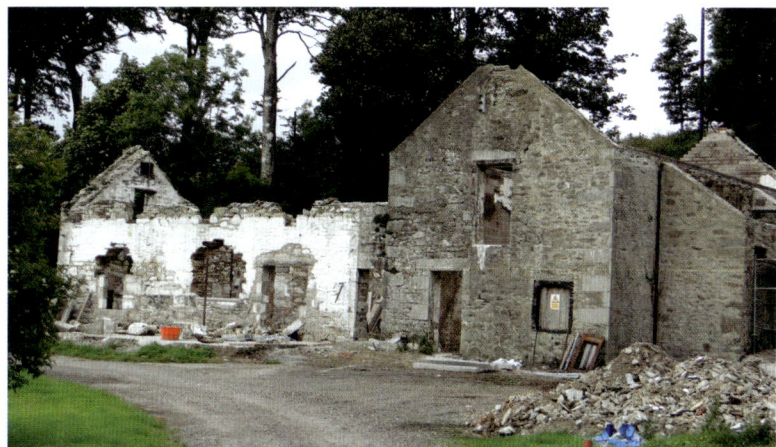

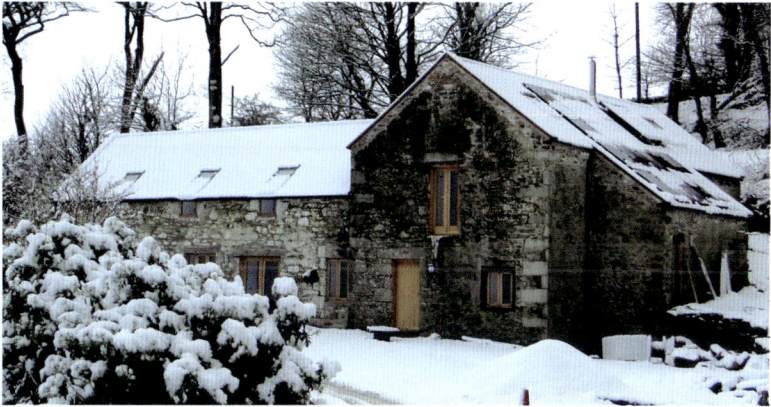

Barholm Mill, along the track from Barholm Castle, before and after restoration by Michael and Sarah Crothers in 2012.

the estate up until the 1970s, and for a while it was the only surviving water-powered sawmill in the country. Sadly, the mill pond was largely washed away in a violent storm in 2022, and its wheel can now no longer function.

In 1987 Patrick Whitford bought Barholm Castle and Barholm House (the new house down the track, not the mansion in Creetown) for £65,000 from Mr and Mrs Hogg, and in 1988 he bought Barholm and Cairnyhill Woods for £19,390 from the Kirkdale Estate Trust. In 1997, having been unable to fulfil his plans for the restoration of Barholm Castle, he put it on the market for £65,000. We bought it in 1999.

This history of Barholm and the Barholm McCullochs may be detailed, even rambling, but it is incomplete; there is much research still to be done to flesh out the details further, and many questions left to answer. I leave that to others. Perhaps a McCulloch descendant will take up the challenge, or a future occupant of Barholm Castle? I wish them the best of luck!

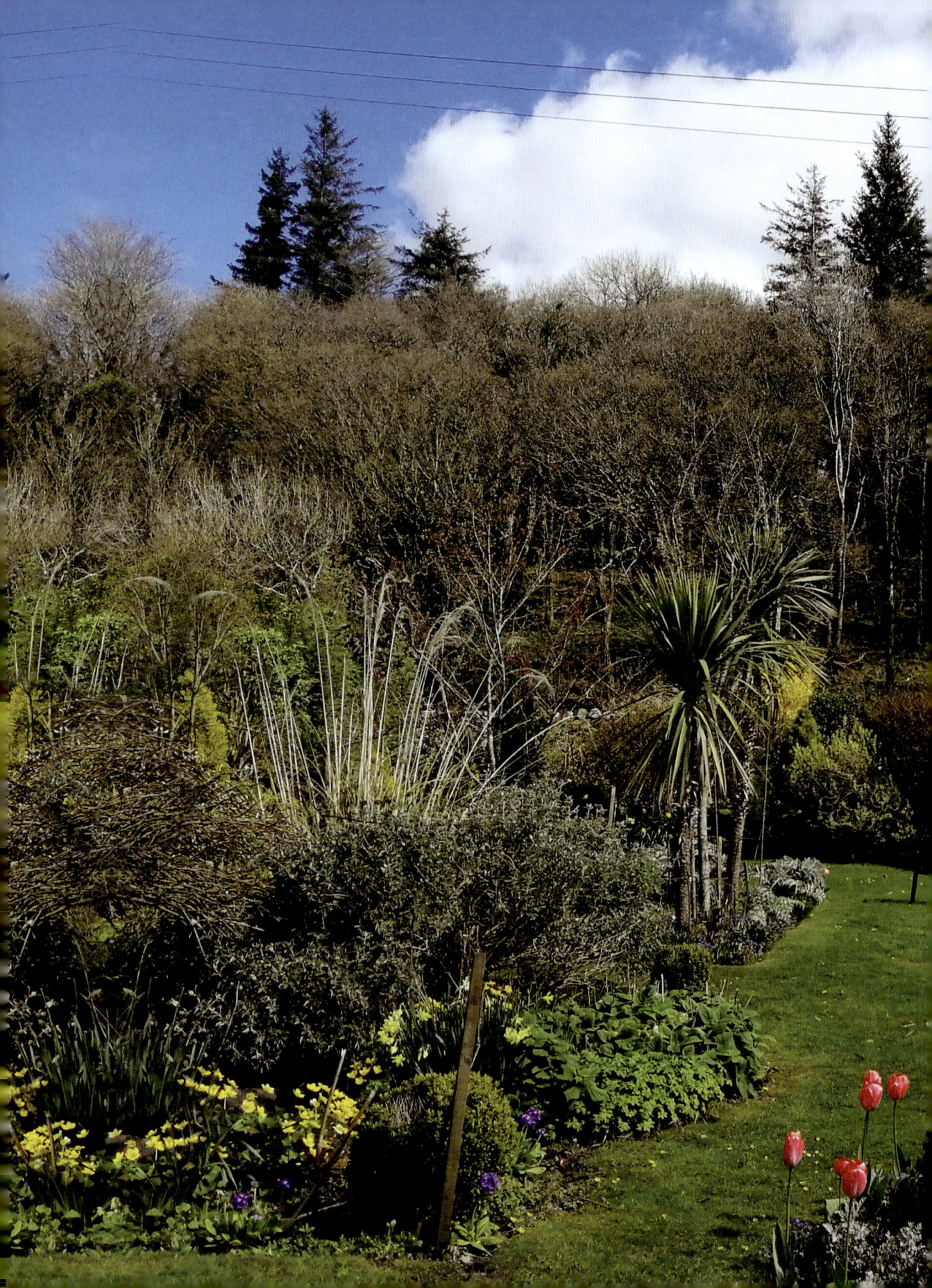

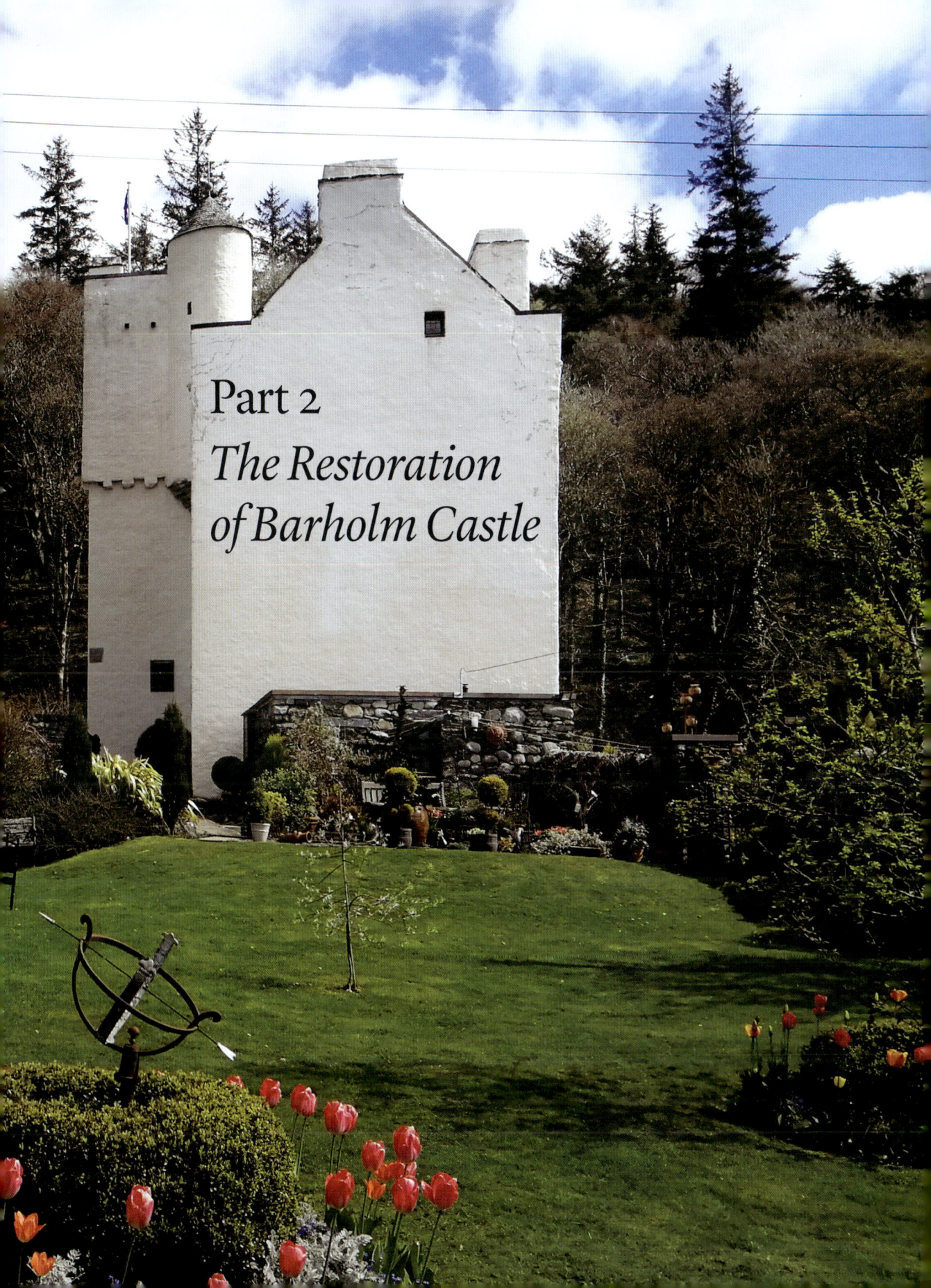

Part 2
The Restoration of Barholm Castle

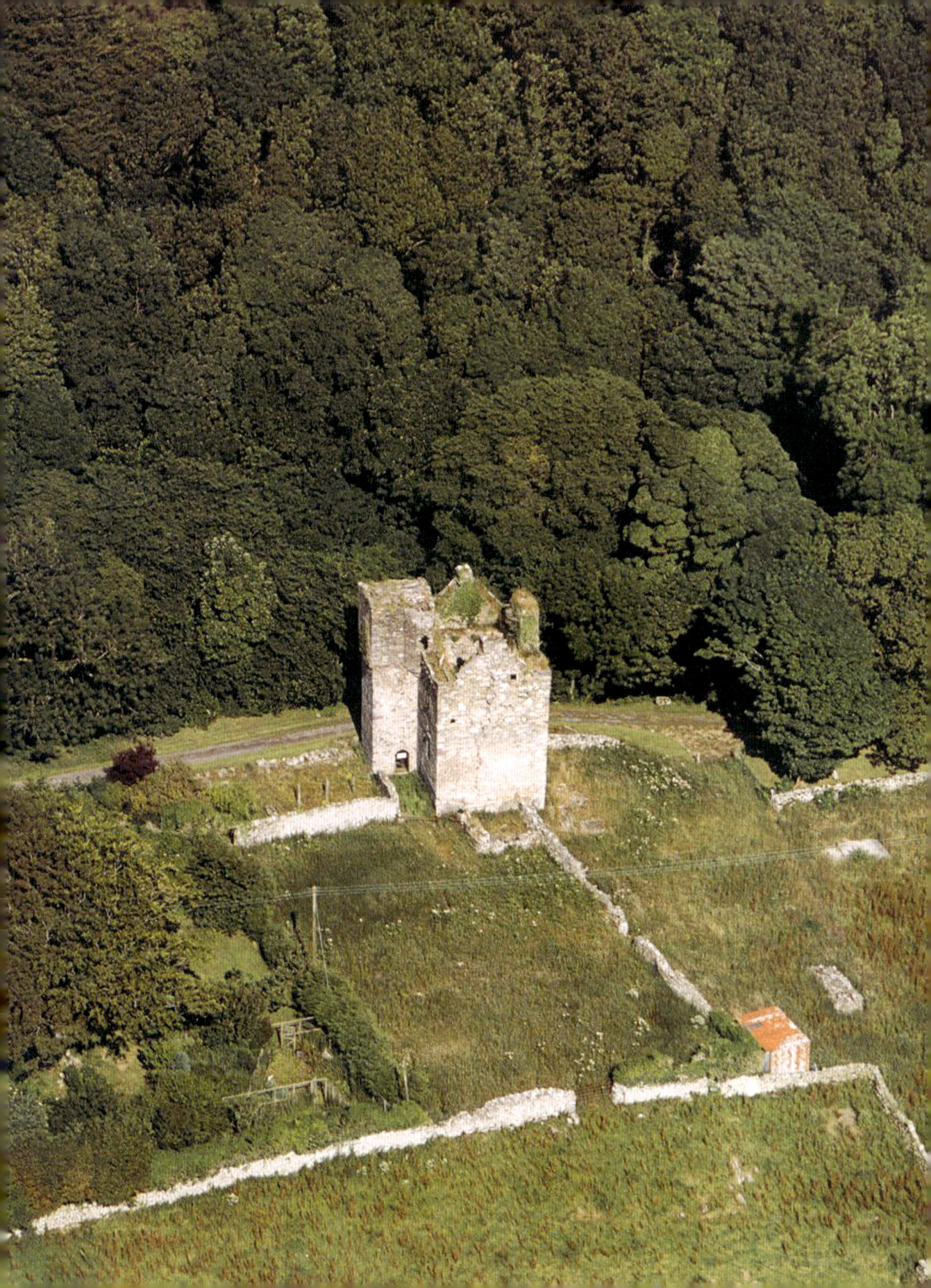

The Restoration of Barholm Castle

For more than twenty-five years before we bought Barholm Castle we had been avid castle visitors. We had also enviously looked at a number of restored properties and fantasised – as one does – about living in one. To fuel this particular fantasy, we decided to look in at the agents handling the sale, just to find out the price. Barholm Castle was priced at a bargain £65,000 in 1997. We could afford that, we thought!

As Colley Cibber pointed out in his play *The Double Gallant* (1707):

Old houses mended
Cost little less than new before they're ended.
(quoted in Stell 1996, p. 102)

Three hundred years later, Cibber's words remain true. In 1997, we commissioned a report from the late Ian Begg, a conservation architect with enormous experience of Scotland's tower houses. He built his own new tower house in Plockton, where he lived for several years. Begg was encouraging. He thought that Barholm would be an excellent property to restore but that it might cost as much as £250,000. This seemed like a great deal of money to us then! Needless to say, costs more than trebled from this initial estimate and, even after twenty-five years, it is unlikely that we could recover all of the financial investment made.

The First Four Years: Before the Work Could Start

Our first attempt to buy Barholm Castle in 1997 was abortive. The owner was selling in order to honour a divorce settlement. He had made detailed and extensive plans to restore Barholm but was never able to realise his dreams. He put such onerous restrictions on the purchase that our solicitor advised us that it would not be wise to proceed. Two years later, having swallowed our disappointment, we were back in Scotland for a family celebration and happened to notice the 'Barholm Castle for sale' advert in another solicitor's window. This time, the sale was to be administered by the Court, more ground was included in the sale and the restrictions had been lifted. We paid £66,000, having carried out no surveys or received any reassurances that the restoration of what was a scheduled monument of national importance would even be allowed. It was a big risk, but we had done as much due diligence as we could and we took a leap of faith, in the belief that the project was both possible and manageable. Many people who have visited since have told us

Opposite.
Aerial photo of Barholm taken in 1999 just after we bought it.

that they, too, viewed the ruin of Barholm Castle and considered buying it, but backed off either because of the seller's original restrictions or when they were unable to receive assurances that a restoration would be viable. Perhaps they were the sensible ones. When we saw Barholm Castle up for sale again in 1999 we decided we would buy Barholm House as well (also a bargain at £65,000), but we were too late – unfortunately, it had already just been sold.

Any major building project is a test of one's resilience, patience and resources. The next four years were full of frustration, as we spent time preparing for the restoration. It was probably fortunate that we did not know that this first phase would be so long, since we might have lost heart completely. The first thing we had to do was to liaise with Historic Scotland (HS). In those days the procedure for applying for a grant was truly byzantine. One had to apply, by post, to be allowed to apply for a grant application form in the first instance. When we visited the HS offices in Edinburgh to speak with an inspector, we were told that the organisation was no longer interested in supporting restorations of roofless tower houses and that there was no point in us expecting to be given a grant. We were, however, eventually allowed to apply for an application form, which was a relief, and an even greater relief when we were finally allowed to apply for a grant. We also had to apply to be allowed to make changes to the scheduled monument (Barholm Castle), and even if we did not succeed in persuading HS to give us a grant, we would still have to follow their directives when carrying out the restoration. The first inspection after we purchased Barholm described it as in a 'perilous condition'. This was alarming, but on the other hand it gave some comfort that we were about to save a building that was in imminent danger of collapse, rather than wilfully transforming a romantic ruin in the Galloway landscape.

We had to approach Major Rainsford Hannay who, as the previous owner of Barholm (before the one who sold it to us), had a right of pre-emption (i.e. right of first refusal) on the property, to ask him to waive this before we could make our purchase. Fortunately for us, he was happy to do so. When he was the owner of Barholm he had commissioned the conservation architect Ian Lindsay to draw up plans for restoration of the castle in 1953 and we asked him why he had not then carried out the proposed scheme. His answer was that it was going to be too expensive. This should have been a warning to us.

Lindsay proposed a low extension to the east side of the tower, plus a freestanding garage. He also sketched in three large windows in the west elevation, which, if achieved, would have made the interiors of the kitchen, great hall and bedroom lighter and brighter, but would have meant a great

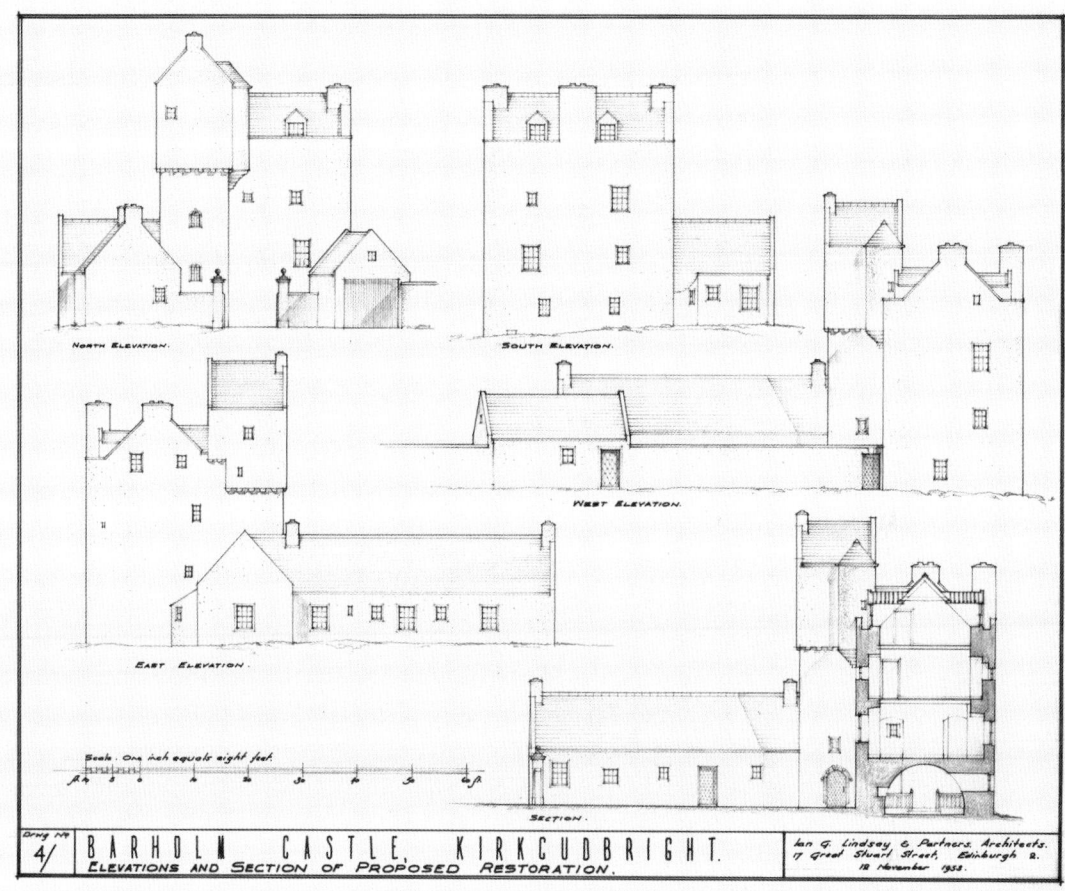

Plans drawn up by Ian Lindsay in 1953.

deal of structural change – inserting large windows in 7-foot-thick walls of random rubble would have been a serious challenge, and would have changed the character of the building.

As a keen early adopter of new technology, my husband John had bought our first digital camera in 1999, an expensive little Sony which stored sixteen photographs on its internal memory stick. These then had to be downloaded, very slowly, to the computer and the memory stick wiped before the next set of photographs could be taken. Extra memory sticks were purchased at great cost. Despite these drawbacks, that camera was a godsend. It allowed us to monitor our images more or less instantly, without having to take spools of shots of unknown quality to a photographic processor and wait for the results. It is difficult now, when every mobile phone has a powerful camera, to remember just how time-consuming, tedious and expensive it used to be to take good photographs. We printed out an A4 sheet with a set of colour images of each of the elevations of Barholm and attached these to our grant application. It added an air of professionalism to the documents, especially in 1999.

In 1999 many small Scottish firms did not have access to the internet or use emails, so the job of communicating with architects, builders, Historic Scotland and others from overseas was more tiresome. I came across an email from me, dated 13 September 2002, thanking the HS officer for her letter offering a grant, for which we had been waiting anxiously for some time. The letter was dated 21 August 2002. I wrote, 'We received both copies yesterday, the original apparently having been sent via Malaysia *en route* to Holland. We confirm that we wish to proceed with the application for grant, on the terms laid out in your letter and enclosures.' Nowadays, we would have received the offer in an email, rather than by post – or no offer at all, as the days of grants for individuals to restore tower houses are gone.

We finally received the building warrant, i.e. approval of the finished building, in January 2006. Barholm Castle was not de-scheduled (removed from the HS list of scheduled monments) until 5 November 2008, although technically this should have happened before any work began on the building. It is a category A listed building. These buildings are defined as 'of special architectural or historical interest which are outstanding examples of a particular period, style or building type'. There are over 3,000 category A listed buildings in Scotland.

The Costs and the Building Process

The indicative overall cost of restoring Barholm Castle, given to us in January 2002 by the quantity surveyor, was £674,960, of which £139,080 was not grant eligible (i.e. related to modern fittings, such as bathrooms and the kitchen), leaving £535,880 of grant eligible costs. We received 33% of the grant eligible costs, a substantial sum nonetheless and immensely helpful to us. The final cost was, of course, much higher, at more than £800,000.

We were expected by Historic Scotland to obtain three tenders for the construction work. This was an initial challenge, as the first two builders we approached declined to tender. Scotland is a small country and does not have an abundance of building firms with the capacity or inclination to undertake a large project involving a building with the challenges that scheduled monument status brings. We finally received one tender for £1,784,255.55, one for £1,344,534.34 and one for £670,785.19. Of course we accepted the last one, which was made by Cumming and Co. of Perth, and they carried out all of the building work for us, using mostly local labour from near Barholm. This was a large and complex project and the bill of quantities, which detailed every last nail to be used, was more than two inches thick.

We had to pay value added tax (VAT) on the building costs. This has long been a source of extreme concern to those concerned with the conservation of historic buildings. Rebuilding, repair and restoration attracts VAT at 20%, whereas new buildings are zero rated. Twenty years on from the restoration of Barholm, when we are all much more aware that the preservation and upgrading of existing buildings is far less damaging to the environment than the building of new ones, it is scarcely credible that the government persists in heavily penalising those who do the former.

There were around fifty people directly involved in the restoration: architects, quantity surveyors, structural engineers, builders, archaeologists, masons, joiners, electricians, plumbers, painters, artists, building control officers, fire officers, Historic Scotland inspectors, solicitors and others. Mostly they made their contributions with professionalism and worked well together. The quality of the work overall was excellent. However, there was some conflict. Our close neighbours, who had just pipped us to the post in buying Barholm House down the track, objected to the restoration from the start. They had doubtless imagined a peaceful rural life at the end of a quiet track. They complained vigorously at every stage of the project, bombarding us with angry emails and causing problems for the workmen. On two occasions the police were called out. I spent many hours patiently and politely replying to raging email messages, which only generated even wilder replies. They stopped, to my surprise, only when I wrote and said that in future I would only respond to messages that were polite and respectful.

When I was interviewed by the BBC for a *Restoration* television series in 2002, I was pressed hard with questions about interior design. I must have been a great disappointment to the researcher, since all I could come up with was the fact that we were going to paint the walls white. I later watched one of the programmes, where the owner of a restored seventeenth-century northern English manor house was flicking through furniture catalogues provided by the television company and choosing rococo and baroque pieces, to the amusement – and obvious contempt – of the design guru. The object of the exercise seemed to my cynical eye to be to entice the woman to make a fool of herself, so that those of superior knowledge and taste might snigger. Thank goodness we were not chosen for the series. A snooty designer might

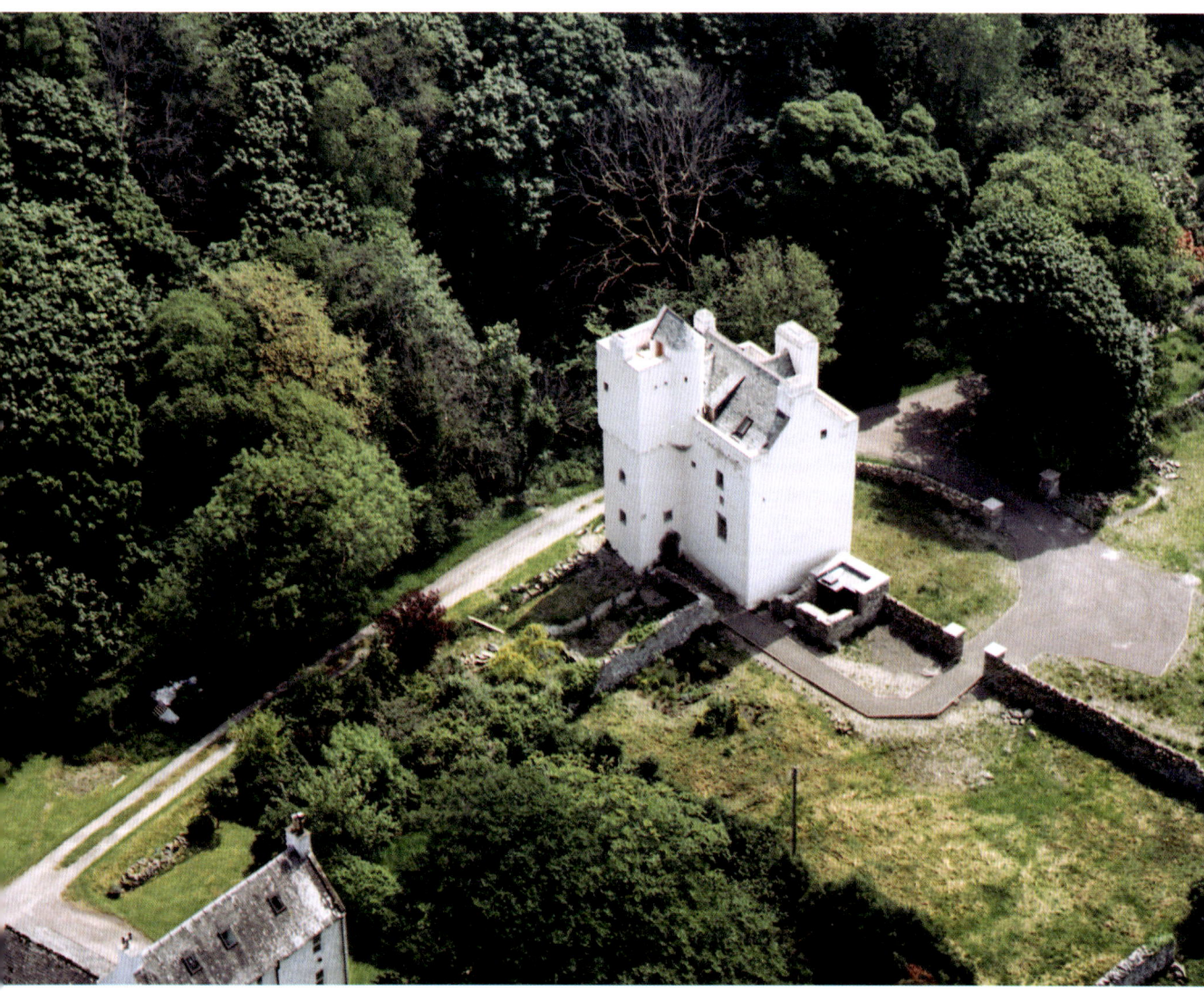

Aerial view of Barholm taken in 2006, not long after the restoration was completed.

have scoffed at our fake Dutch antiques and our reproduction four-poster beds. Besides, design planning takes the fun out of serendipitous finds in unusual places.

We did appear on a couple of television programmes after the restoration. We featured in an episode of *Extreme Homes*, alongside a New York home inspired by a diamond ring, a house in Mexico shaped like a shark and a South African adobe village painted in colourful symbols. We were also on *Homes by the Sea* when it featured the Galloway coast. A few years ago friends were flying to Oman and turned on the in-flight entertainment system. The first thing they saw on the screen was me and John in Barholm, appearing in *Extreme Homes*!

We intended to use Barholm for holiday letting and so planned to furnish and decorate it accordingly. I had great fun in Dutch snuffel markets, snuffling out bargains. Second-hand goods sales have always been enormously popular there, with a huge range of very cheap goods on offer. I was looking mainly for pewter, copper and old rugs, which I often found for excellent prices. There must have been a fashion for pewter jugs and plates in the Netherlands in the 1970s, as so many are for sale among other items from that era. A set of pewter measuring jugs cost 5 euros. A vast and extremely heavy copper cauldron pot in excellent condition was 20 euros. A large, old copper jelly mould was only 1 euro. In addition, I picked up all sorts of unusual odds and ends, such as old wooden *speculaaspop* (gingerbread man) moulds and several intricately carved models of spinning wheels. A large wolfskin rug originally bought for the turret bedroom was only 10 euros, although eventually it had to go, as unfortunately it turned out not to have been well cured. The little rugs that sit on the spiral stairs all came from Dutch markets, usually at a cost of 5 euros each. The framed Dutch tiles lining the staircase were also an average price of 5 euros and the blue and white tiles laid on boards to line the fireplaces come from my collection of tile oddments at a euro or two each.

I recently came across an upbeat Christmas letter that I sent to a friend in December 2002:

> Things are looking good for our castle. Historic Scotland has finally agreed to our restoration plans and, moreover, has offered us a good grant. The applications for planning permission, building warrant and de-scheduling have been submitted, and we are hoping that work can begin next spring, although delays do seem to be built in everywhere along the line. The optimistic plan is to try to get the roof on next summer and possibly finish work the summer after. We'll see. At any rate, five years after we originally saw Barholm, we finally feel we are making real progress towards restoring it.

Work started in June 2003, or at least that was when the contractors first came on site. We were thrilled to see evidence of digging when we visited in June that year, but the only outcome of that work was the insertion of a septic tank (which I must admit has served us well over the twenty years since then, without ever needing emptying or flushing out). Actual building work did not start until late autumn, when the scaffolding started to be erected.

During the thirty months of the restoration, we commuted to Scotland during school holidays, taking the overnight ferry from Amsterdam to Newcastle and driving to Barholm via Dumfries, where John's parents lived. When we visited the building site surrounding Barholm Castle, to have meetings with architects and builders, our daughter Rose, who had not long learned to read independently, would sit in the car and read the Harry Potter books. I think our visits to Barholm got her through the entire series. She also learned to solve Sudokus, at that time a new craze in the world of puzzles. She still loves a Sudoku challenge, a legacy of Barholm for her.

Archaeology

An archaeological watching brief was part of the grant stipulation from Historic Scotland, which we had to pay for. Kirkdale Archaeology was awarded the contract and produced a thorough set of photographs of the building's exterior and interior before any work started. They also produced a lengthy report, which has been very helpful in tracing the built history of Barholm Castle and helping us to understand some of the quirks in the building.

The setbacks

One day in 2003 when I was at home in the Netherlands I answered the phone to Peter Drummond, our architect. 'Are you sitting down, Janet?' he asked. Anticipating bad news, I immediately sat down, and listened in horror as he told me that it had been discovered that the south wall had a 'belly' which meant that it would have to be taken down; otherwise it would collapse. All of it had to come down, to ground level, and be completely rebuilt. Worse still, although the wall – an average of 7 feet thick and 20 feet high by 20 feet wide – was made of random rubble, like a gigantic stone dyke, the inspector had decreed that each stone had to be numbered and put back exactly where it came from. Then it was all to be covered up with a coat of harling, so that none of the stones would ever be seen. Fortunately, Peter, who had previously worked for HS, was able to convince the inspector that this was an unnecessary extravagance that would cost us dearly in both time and money. The

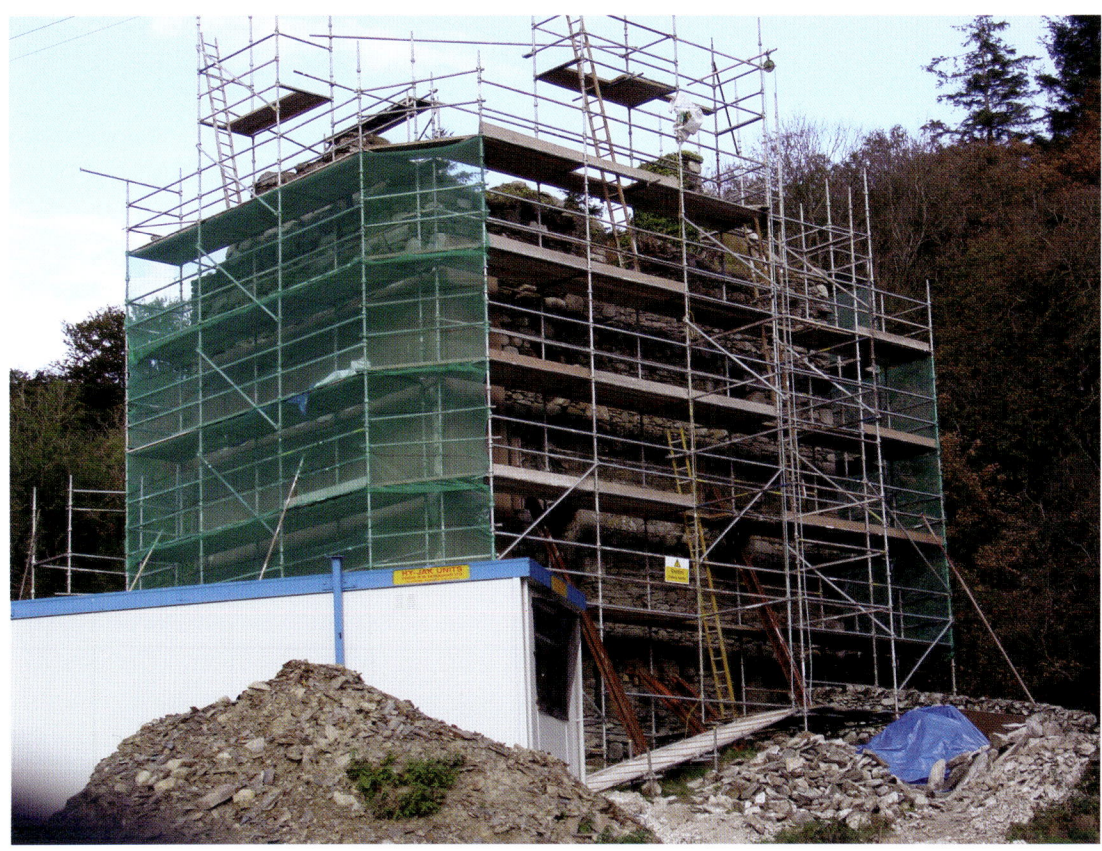

Taking down the south wall; its 'belly' was making the whole building unstable.

apparent cause of the bellying was the giant chimney, which was inserted during the second phase of building and not properly keyed into the stonework of the wall. A botched job by an incompetent builder; they existed in the sixteenth and seventeenth centuries too. The laird who carried out this extra work must have felt that the perceived benefit in status was worth it; he took a risk that paid off for him, but ended up causing trouble in the twenty-first century.

For a year after the agreed completion date we were hit by constant disappointments, as firmly promised deadlines approached and were successively not met. At first, we shrugged our shoulders and thought, oh well, what's an extra month in the grand scheme of things. But as the financial implications of the delays sank in, and after we had to cancel the opening party (timed for nearly eight months after the completion date and in the middle of our summer holiday), we began to acquire a protective coat of cynicism, accompanied by real anxiety about the amount of extra money which the delays were costing us. As with childbirth, my memory has protectively blanked out much of the pain and distress we experienced.

Looking back through the reams of letters and emails sent and received during the restoration, I am astonished that we managed to cope. I was working in two jobs, as an Open University tutor and also heading a department in an American University in the Netherlands. John was working across the Hague and Munich in a demanding job, commuting between the two cities most weeks. Fortunately, our daughter Rose was a very undemanding and self-sufficient child. As an adult, perhaps unsurprisingly, she is not much interested in castles or historic buildings!

The Interior

At Barholm Castle, cooking would have been done either in a separate outhouse or over the fire in the great hall, which by the time of abandonment had been partially blocked off to save fuel, as had the other fireplaces in the house. When coal became more readily available as fuel for heating and cooking in the seventeenth century, fireplaces were reduced in size and adapted to burn the 'devil's excrement'. It was clear that the fireplace in our bedroom had been radically reduced in size.

Historic Scotland insisted that the fireplaces should remain at their smaller size, as part of their policy of preserving buildings at the time of their last occupation. Accordingly, we were made to plaster over an aumbrey, or recess, discovered on the second floor, which had been roughly filled with stones, as this was evidence of how it had been at the last occupation. The plastering happened while we were away, so to our annoyance and frustration we have never seen that aumbrey. However, the policy was not applied consistently. As can be seen from the photograph of the interior of the west wall, an opening there had also been filled with stones. We were allowed to remove these and discovered what had been a buffet or cupboard set into the wall. We also discovered, after the restoration was complete, RCAHMS photographs showing that the fireplace in the great hall had been partially blocked up. It is lucky that the HS inspection team had not looked at these photographs, otherwise we might have been made to have a small fireplace in the great hall, which would have looked out of proportion and out of keeping with the character of the room. Stories in the conservation world – some of which may be urban legends – tell of owners of derelict eighteenth-century properties being made to retain ugly and wholly inappropriate 1960s

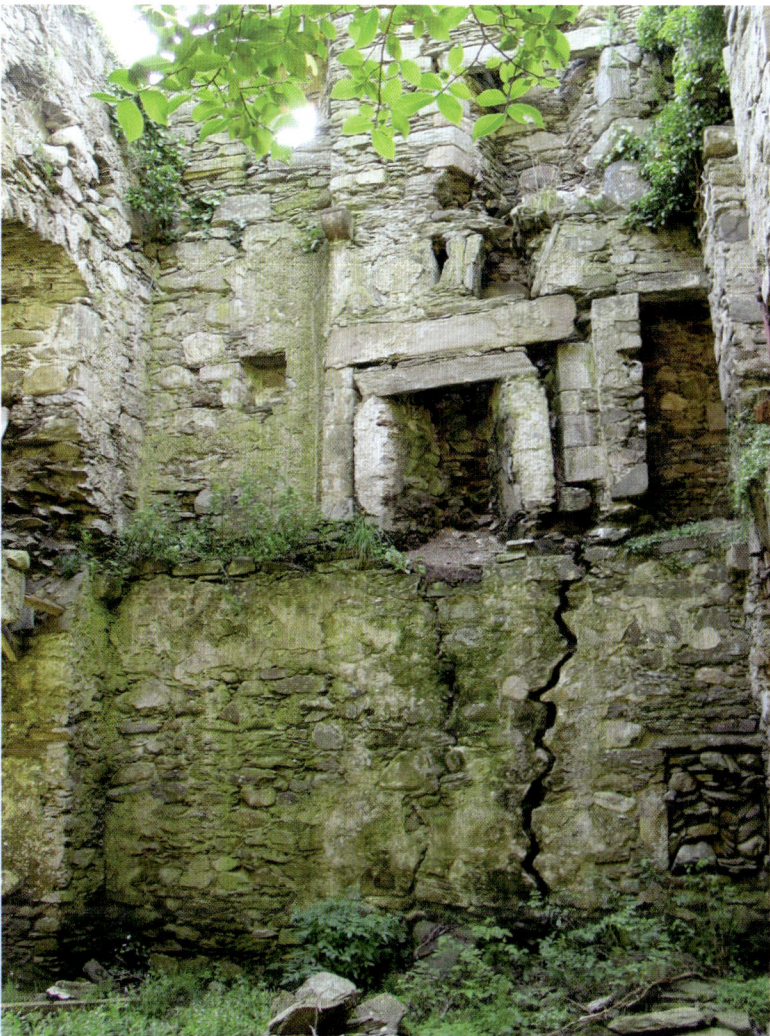

The interior of the west wall, showing the fireplace in the master bedroom, clearly having been reduced in size by placing additional large stones at the sides which hold up a new, lower lintel. This photograph also shows the aumbrey, or cupboard, in the great hall, which was packed with stones (bottom right).

accretions when undertaking renovations, in order to conform to the 'last evidence of occupation' rule.

Historic Environment Scotland (renamed in 2015) must now grapple with the contradiction between its strict policies on the use of period materials in conservation and the need for energy efficiency in our buildings. Double-glazing technology is already sophisticated enough to be scarcely noticeable in the windows of heritage buildings. Not only does its use conserve energy, helping to keep the building warm and free from draughts, it also considerably lowers the fuel bills of owners who may be struggling to keep up with maintenance costs. At least HES agreed – when I was a Board member of the

John, architect Peter Drummond and the site supervisor inspecting the west wall after a scaffolding tower had been erected.

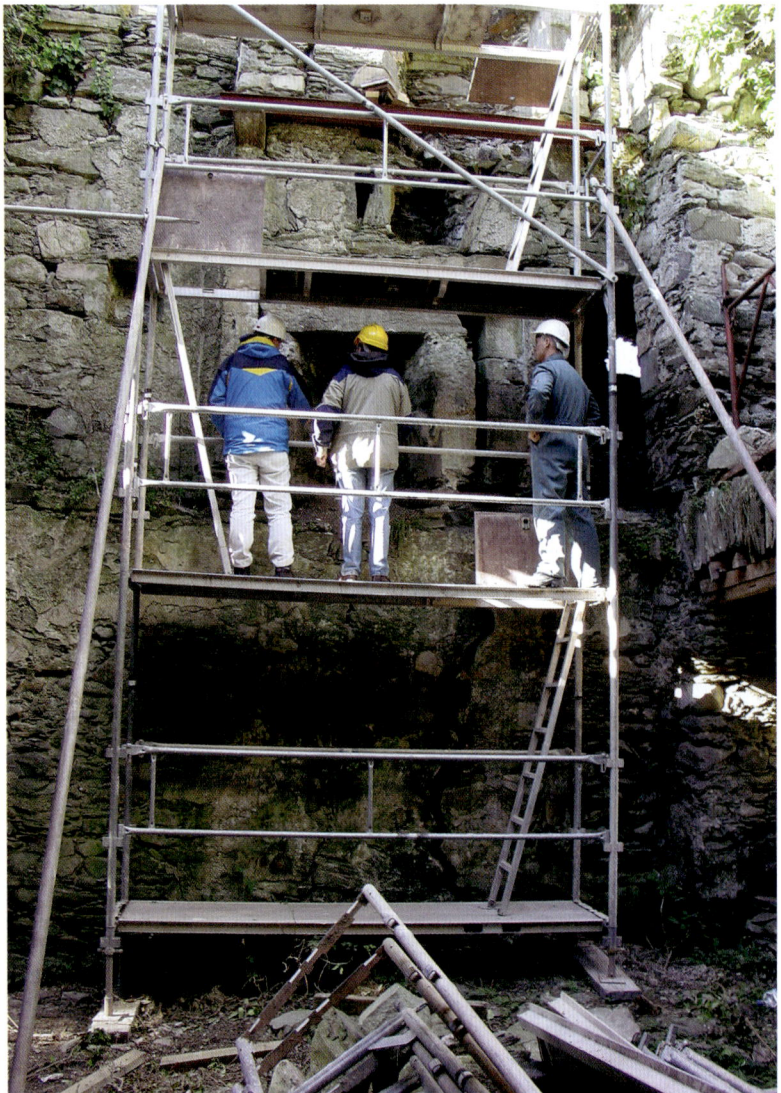

organisation – to allow solar panels to be installed on the back of one of the buildings of Edinburgh Castle, out of sight of the public.

We are often asked about how cold it is living in a Scottish castle. In fact, it is very cosy indeed. Gas-fired underfloor central heating was installed beneath the flagstones of the kitchen and great hall, with radiators in the bedrooms and bathrooms. We paid the farmers for wayleave to bring the mains gas pipe from the A75 road across their field to the castle and have enjoyed reasonably priced central heating ever since. Most sixteenth-century towers in rural areas, however, have to rely on oil, bottled gas or electricity for heating unless they have ground-source heat pumps or some other source of renewable energy, as is the case at Rusco Tower near Gatehouse of Fleet.

The spiral staircase

The staircase – spiral, turnpike or wheel – was almost intact when we bought Barholm. One or two of the steps needed to be repaired, but they were not badly worn. That is because the stone is hardwearing greywacke and not soft sandstone. As one climbs the stairs, the undersides of the almost perfectly regular pie-slice-shaped pieces of stone loom above, laid at precise angles round the newel post are an amazing piece of stone engineering, almost a work of art. I often think of the men who cut, lifted and positioned these huge stones. What strength, skill and perseverance they must have had! There is a mathematical formula for working out the angle of spiral stairs ($\pi \times$ radius $\times$ angle of rotation $/ 180$), but it is certain that the builders did not use this in any formal way. They may have used a template, however, to cut the treads to the exact size. There is a date scratched into the exterior of the stair tower: 1375. It seems remarkably early, until one realises that the

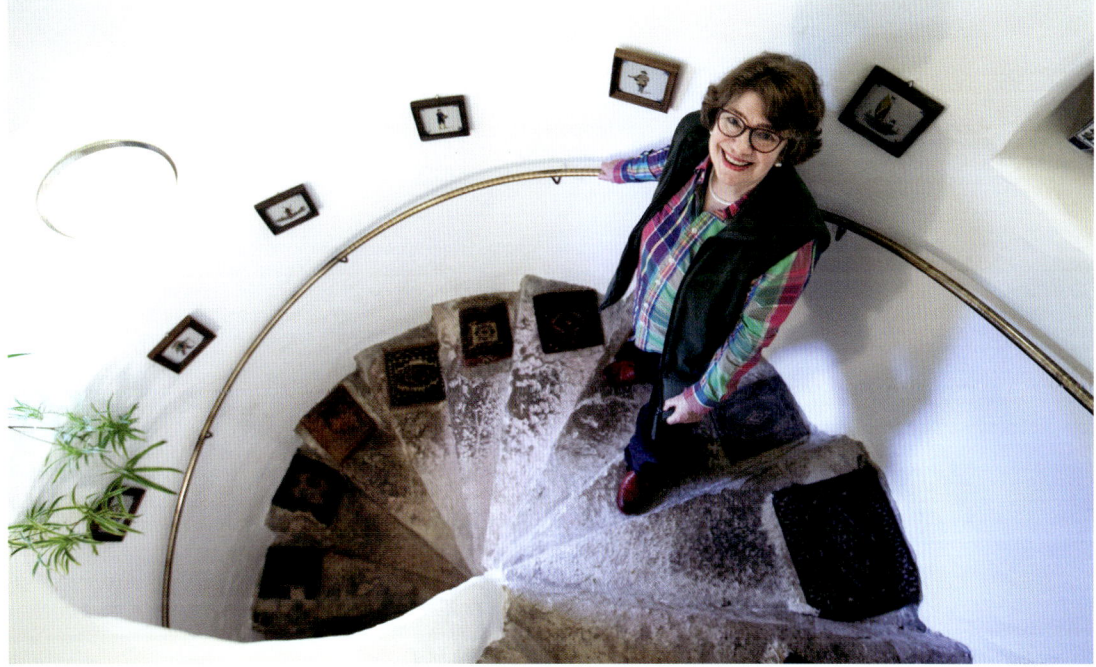

The spiral, or wheel staircase, a work of expert stone engineering.

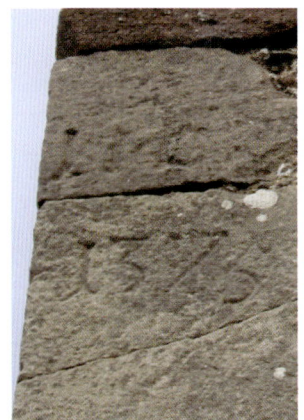

The date '1375' carved into the corner stone of the stair tower; the mason probably intended to write 1575. Above the date is what looks like a mason's mark, but it is very unclear.

masons would have been illiterate and not quite able to distinguish between the numerals 3 and 5. Thus 1573 (or 1575) is much more likely.

Tony Morrell, a talented blacksmith from Ayrshire, made and installed the elegant bronze handrail which has served us well. It looks great and gives a secure support for climbing and descending the stairs. Much better than a wobbly rope!

The kitchen

The ground floor was called the 'dungeon' by generations of local children who frightened each other while exploring the dark depths of the vaulted cellar in the ruined castle. It has been transformed into a cosy but spacious kitchen with underfloor heating. Near the middle of the room was an outcrop of raised stone, which the builders advised should be cut down to the level of the rest of the floor in order to make it flat and allow the even laying of underfloor heating pipes. The Historic Scotland inspector initially said no: the raised stone had to stay according to the policy of conservation. This was a terrible blow, as the kitchen, with its low vaulted ceiling, would become absolutely tiny and probably unusable as a result. I took the desperate step of writing the inspector a personal letter:

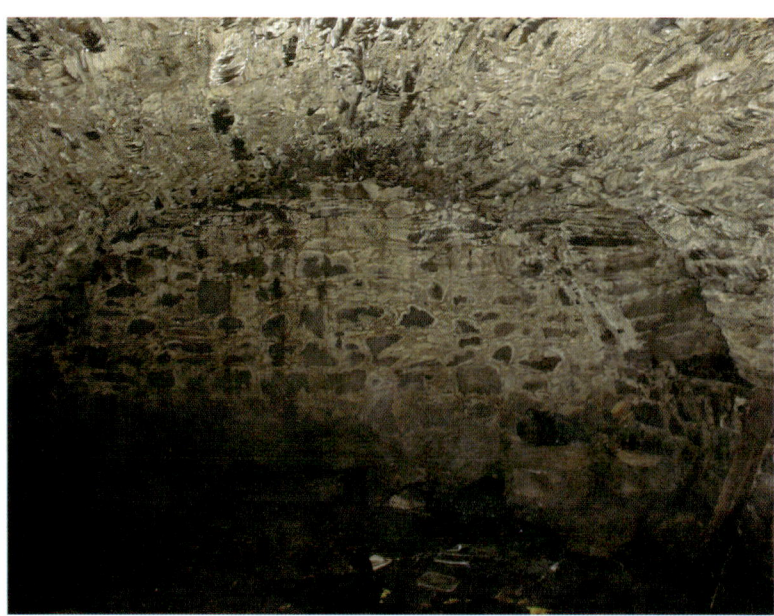

The kitchen before work began. Light came from two small windows in the south side.

THE RESTORATION OF BARHOLM CASTLE

The problem is that if we cannot level the floor somewhat, then there is going to be so little headroom within what will be our kitchen that it will scarcely function as a room. The alternative – large rocky outcrops in the centre of the floor – would be a terrible hazard. While we are completely committed to conserving historical detail within the castle and accept that there must be many areas where consistently applying the principles of conservation will have incurred considerable cost or will cause us

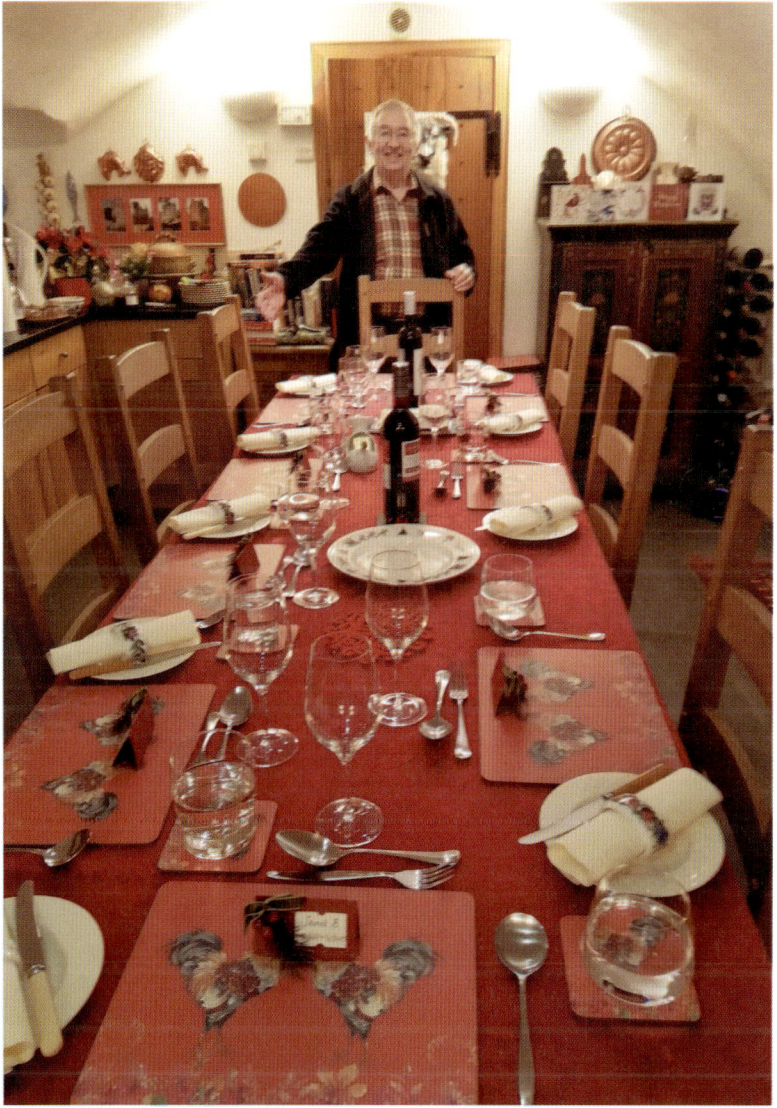

The kitchen after the work was complete, showing the table set for a pre-Christmas meal. Behind the door is a small utility area and a toilet.

inconvenience, we do need to be able to use Barholm as our home. If HS will allow us the installation of electric wiring, piped water, gas central heating and the installation of a modern kitchen and bathrooms within the castle – some of which will surely necessitate a compromise with the principles of historic conservation – can we not equally have a kitchen with a ceiling high enough to stand upright in with ease and comfort? We would be most grateful if you will re-consider this issue and allow us to take enough stone off the isolated high spots with the basement floor area to allow sufficient room to stand and walk around in the kitchen.

To her credit, and our relief, the inspector agreed, and we were able to go ahead with making the kitchen a reasonable size.

The great hall

The great hall was a gloomy space when we first entered it. A barn owl startled us by soaring noiselessly away. When we looked up, we could see the sky above, but it took a few seconds for our eyes to adjust to the lack of light inside and to take in the ruined walls, covered in algae. An empty fireplace hung above our heads on the west wall – now in our bedroom. The huge

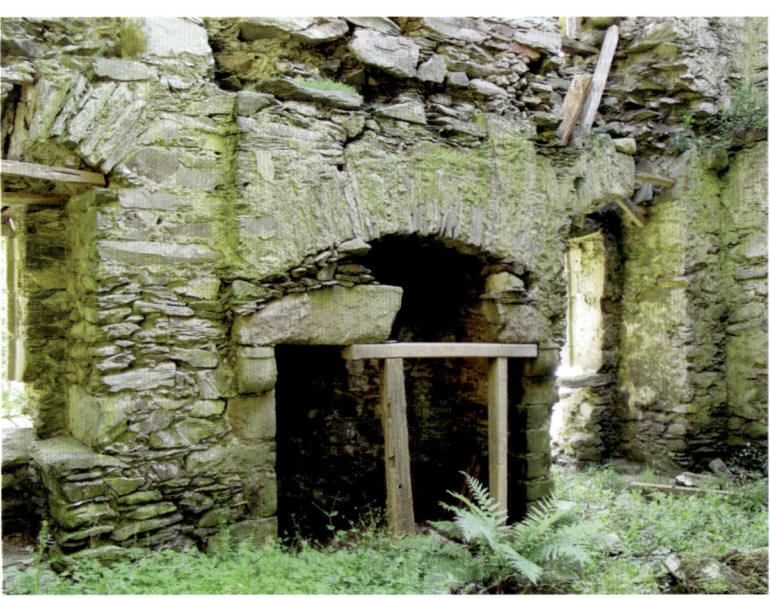

The great hall as we first saw it, damp and gloomy, with its large fireplace; owl pellets littered the floor.

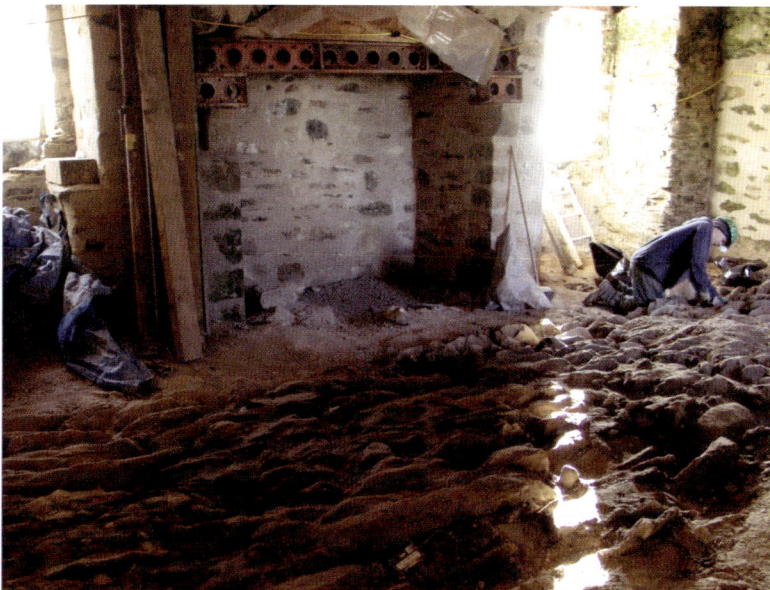

The floor of the great hall is the roof of the vaulted cellar and was in a rough state. Underfloor heating was laid under the slate slabs which were laid on the floor.

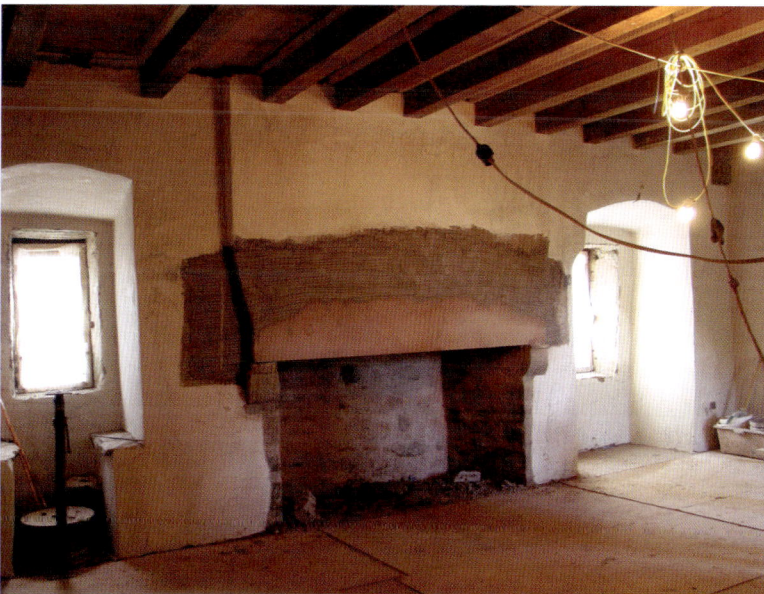

The new lintel in place over the great hall fireplace, and the underfloor heating in place.

great hall fireplace had half of its broken lintel held up by a wooden prop, and the relieving arch was exposed. We could stand inside the fireplace and look up, craning our necks to see the distant light at the top of the chimney.

Over the course of the restoration, the great hall was transformed into a bright, colourful room, with a beautiful painted ceiling. We do not know

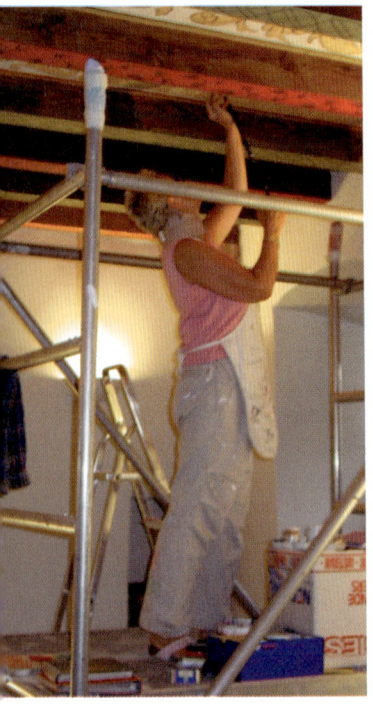

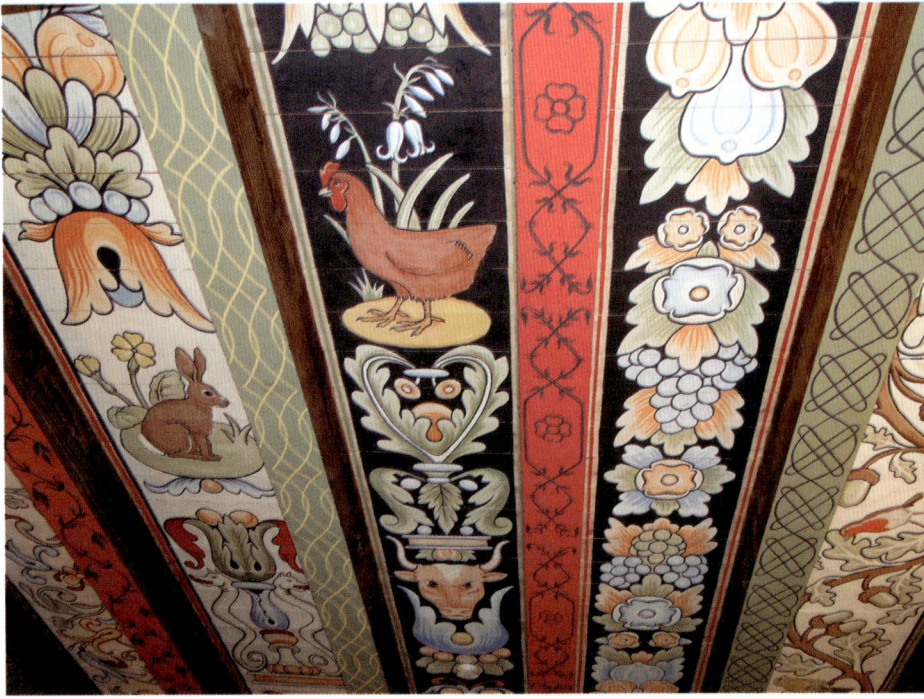

Above left.
Jennifer Merredew painting the ceiling of the great hall with motifs copied from the ceiling of Crathes Castle and images of local nature and wildlife. She painted while standing on a platform, wearing a surgical collar to protect her neck.

Above right.
A detail from the finished ceiling of the great hall.

Right.
The great hall transformed.

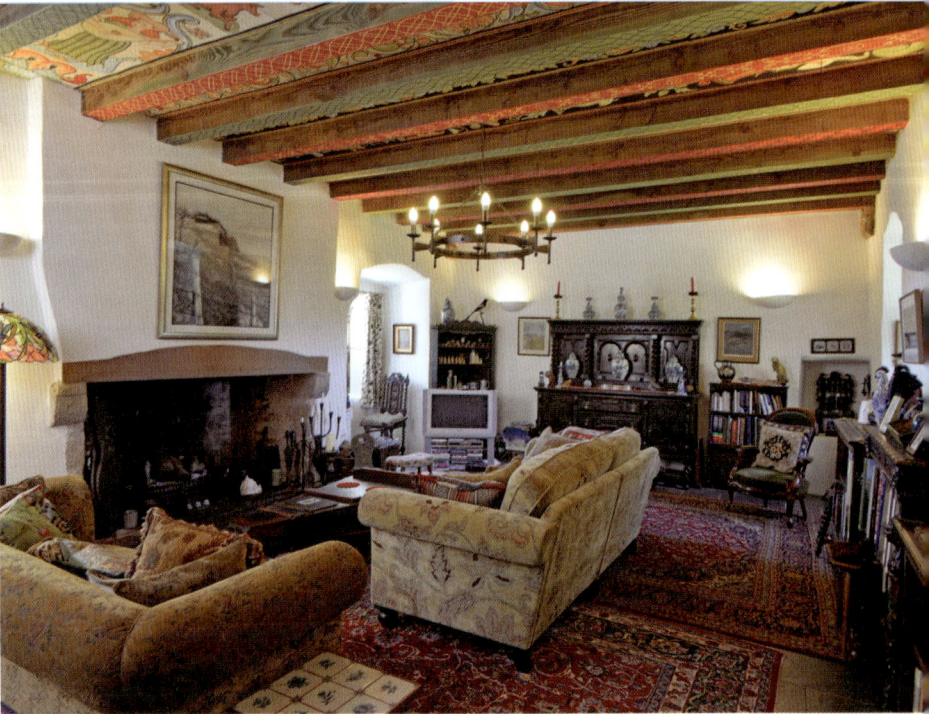

THE RESTORATION OF BARHOLM CASTLE

whether the original ceiling had been painted, as all of the wood had rotted away centuries before, but decorated ceilings were common in tower houses and it seemed a good tradition to carry on. The great hall would have been a dark space in the sixteenth century, but almost certainly one illuminated with rich colours provided by rugs, paintings and wall hangings.

The master bedroom

The window in the master bedroom is larger than those in the great hall and does not have shutters. This is because the shape and size of the window was indeterminate and there was no evidence of its original features, so we were allowed to fill all of the space that existed with window. This floor has a little corridor in the width of the wall giving access to both chambers through a double door. The second chamber would probably also have been a bedroom, but we use it as a large bathroom/dressing room, with a small laundry room off the corridor.

The bed came from a range of the now-defunct Woodcarvers' Guild

From the corridor that leads into the master bedroom, we could look down into the great hall – there was no floor to impede our view. We could also look up to the sky and see the walls of the room above, with the fireplace set into the gallery's west wall.

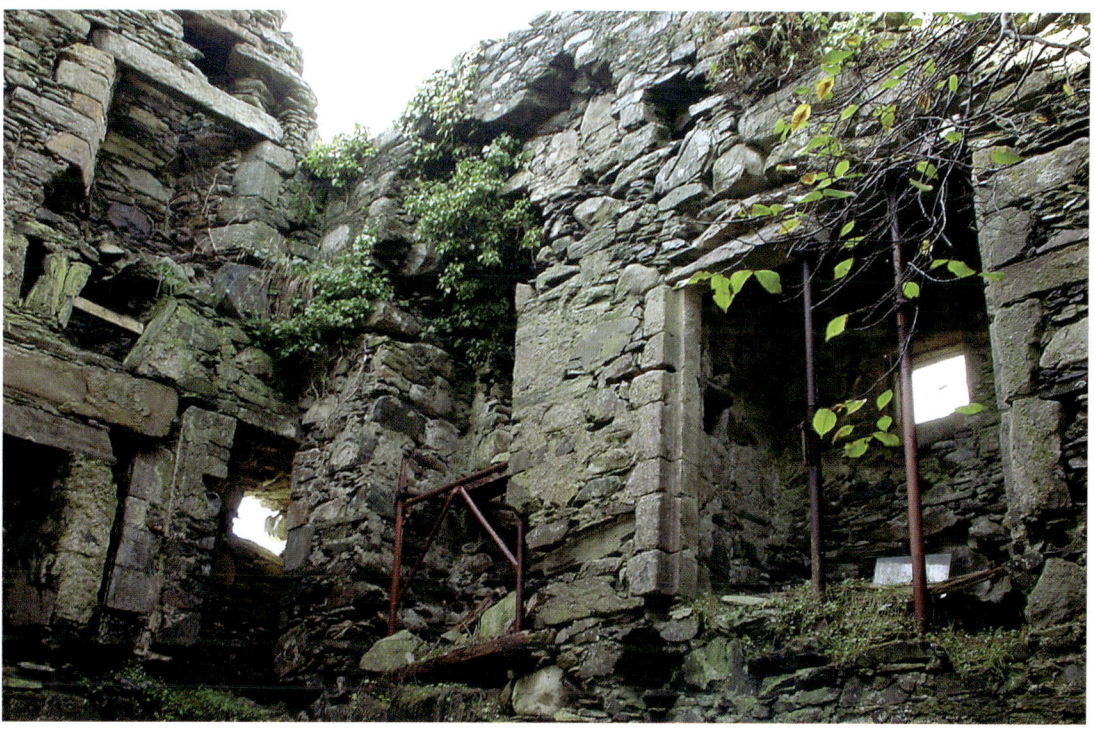

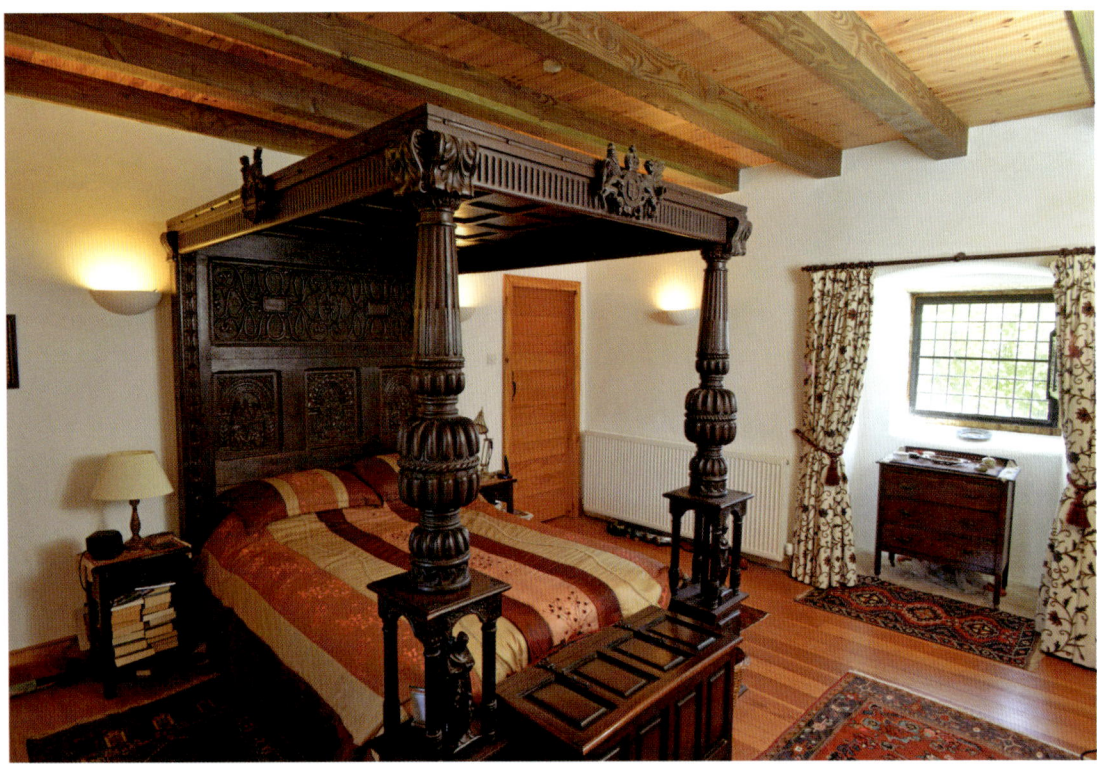

Above.
The master bedroom with its reproduction four-poster bed.

Right.
The bathroom next to the master bedroom; this chamber would have been a bedroom originally.

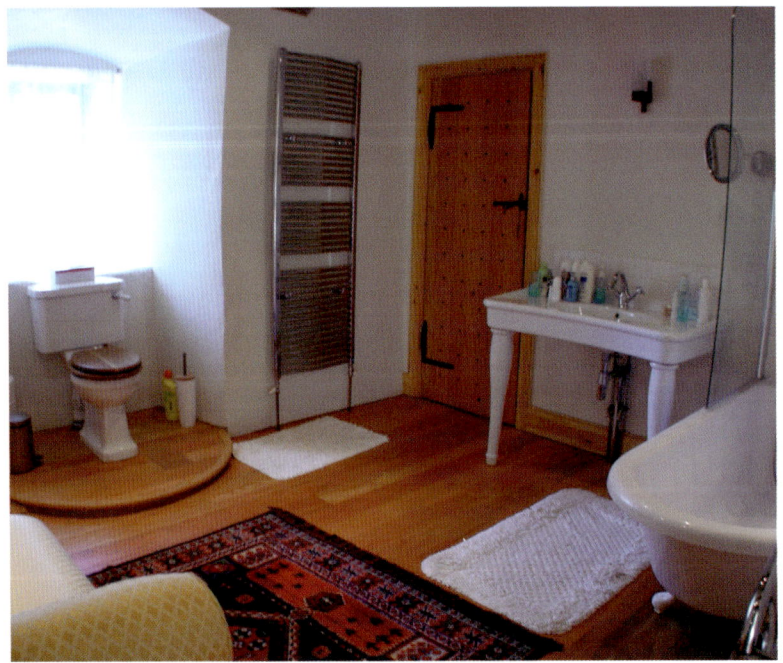

cooperative – we chose the 'Jacobean'. It is a copy of a sixteenth-century original, carved by machine using laser-guided techniques. It came in multiple pieces, of course; four-poster beds have always been 'flatpack', as not even the grandest house has a front door wide enough to accommodate a four-poster bed.

The gallery

The third floor was a long gallery – just an open space with one small window and a small fireplace. In it we have made two bedrooms and two shower rooms (one en suite), accessed by a corridor. In the corridor is a little door, a few feet above ground level, which gives access to the north wall walk. We use the wall walk only for access to clean out the spouties. I imagine that it was the same in the sixteenth century, although it is one area where there are good views of the surrounding countryside and a clear line of sight for

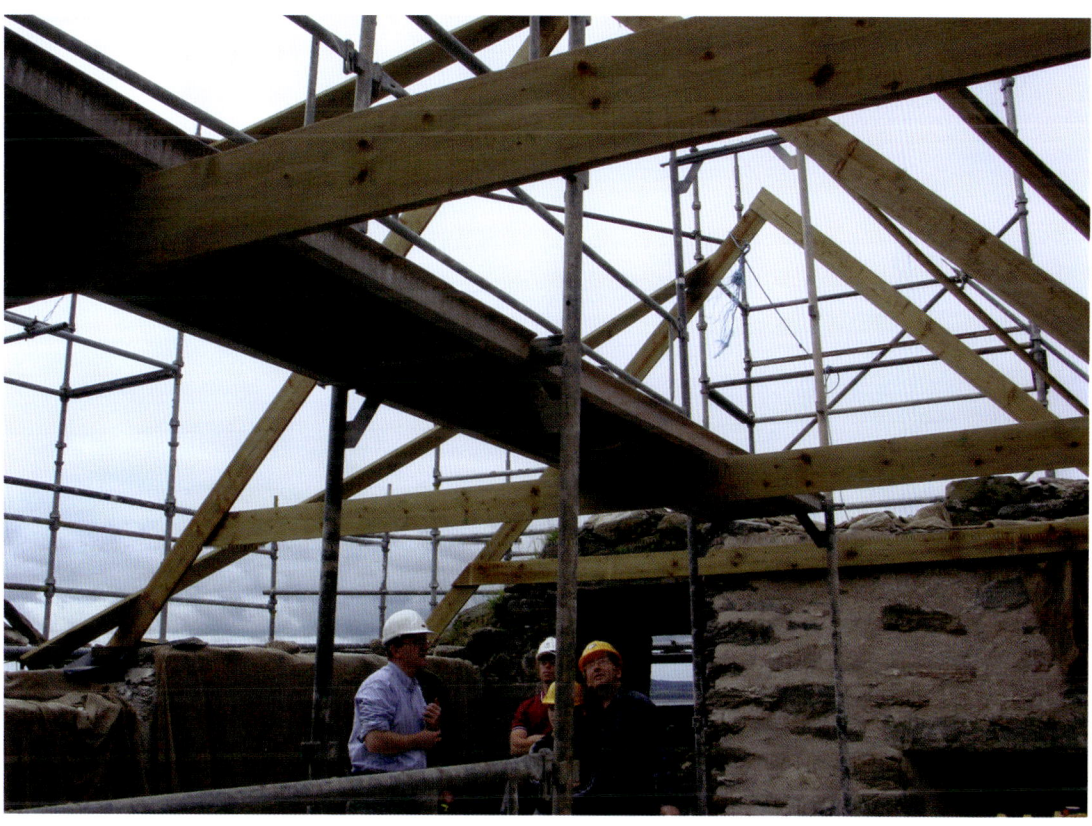

Fitting the roof beams.

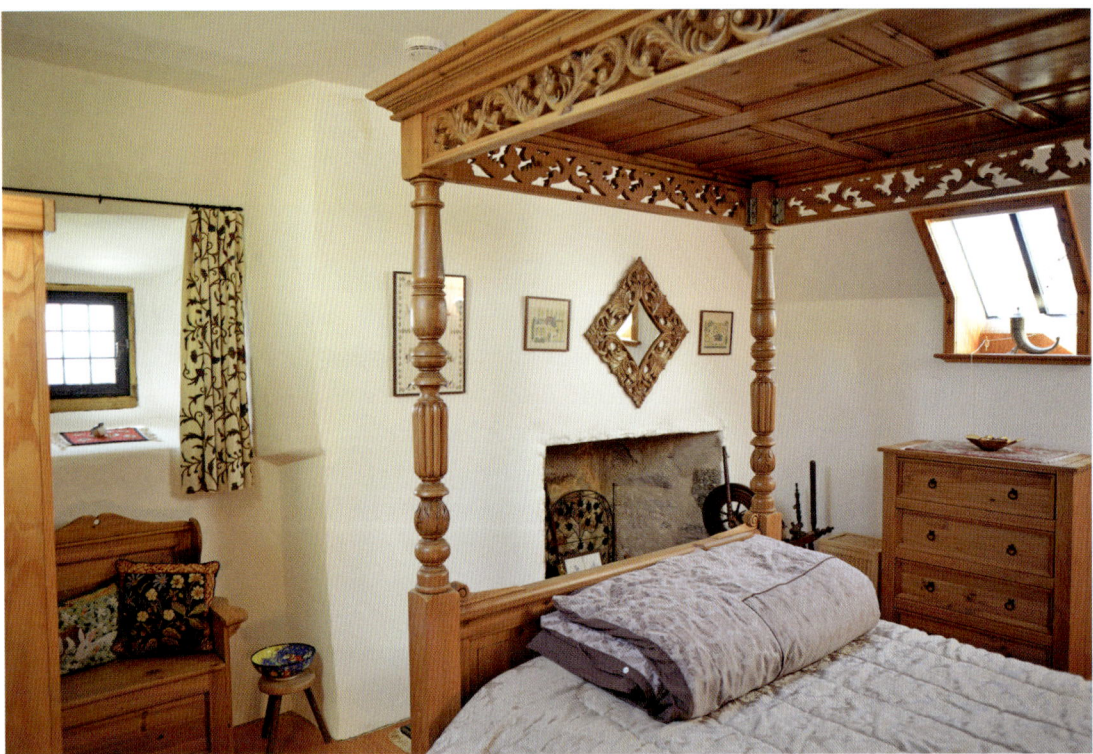

The guest room.

anyone approaching the castle door. Stories of pouring boiling oil over one's enemies can have no basis in fact – how would one even manage to get a tub of boiling oil up the stairs? There may have been dormer windows on one or both sides put in during the second phase of building but there is not enough evidence to be sure. HS required us to fit special 'conservation' skylights on this level; simple Velux windows would have been much easier for us, but the conservation lights may look better in a Victorian kind of way. They are certainly not 'authentic'.

John Knox's room

John Knox's room is accessed by a tiny stone spiral staircase in the corbelled-out turret at the top corner of the stair tower. It has a stone vaulted ceiling, above which is the parapet of the tower. It is relatively unusual to have a vaulted chamber at the top of a tower house. The reason may have been to support a viewing platform with a stone floor, on which a fire beacon could be lit. Barholm is within a direct line of sight of the Isle of Man, an important centre for illegal contraband. Strong beer and spirits were commonly smuggled to avoid taxes; the light from Barholm Castle may have signalled to ships, although it would seem rather brazen and foolhardy to have constructed the top of the tower specifically for illegal purposes. The roof may have been constructed out of stone simply because it was a more accessible material and there was a plentiful supply of cheap labour available to build a vaulted ceiling.

The bed was made for us by the blacksmith at Drumlanrig Castle – our only stipulation was that it had to be made in small pieces, so that it could be carried up those narrow stairs. I tried to buy some wooden furniture from a local antique dealer, but the pieces were too big to be carried up the staircase.

John Knox's room remains as we first furnished it, but no one has ever slept there since we moved in, and it is a room I generally enter only when showing visitors around. Because of its vaulted ceilings, fireplace and radiator, there is virtually no wall space for bookcases or a desk, so it remains a high-level curiosity, not much use for present-day needs. However, if John Knox really did stay there, by sixteenth-century standards he was probably very comfortable indeed, with a nice little fireplace and two windows looking east and north. There is no garderobe toilet, so I assume slops would just have to be tossed out of the window or collected by a servant.

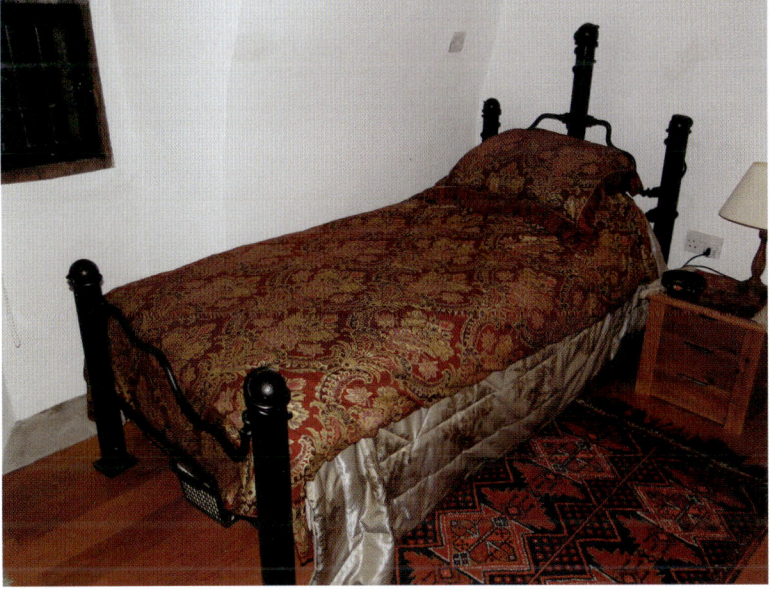

The bed in John Knox's room, made for us by the blacksmith at Drumlanrig Castle.

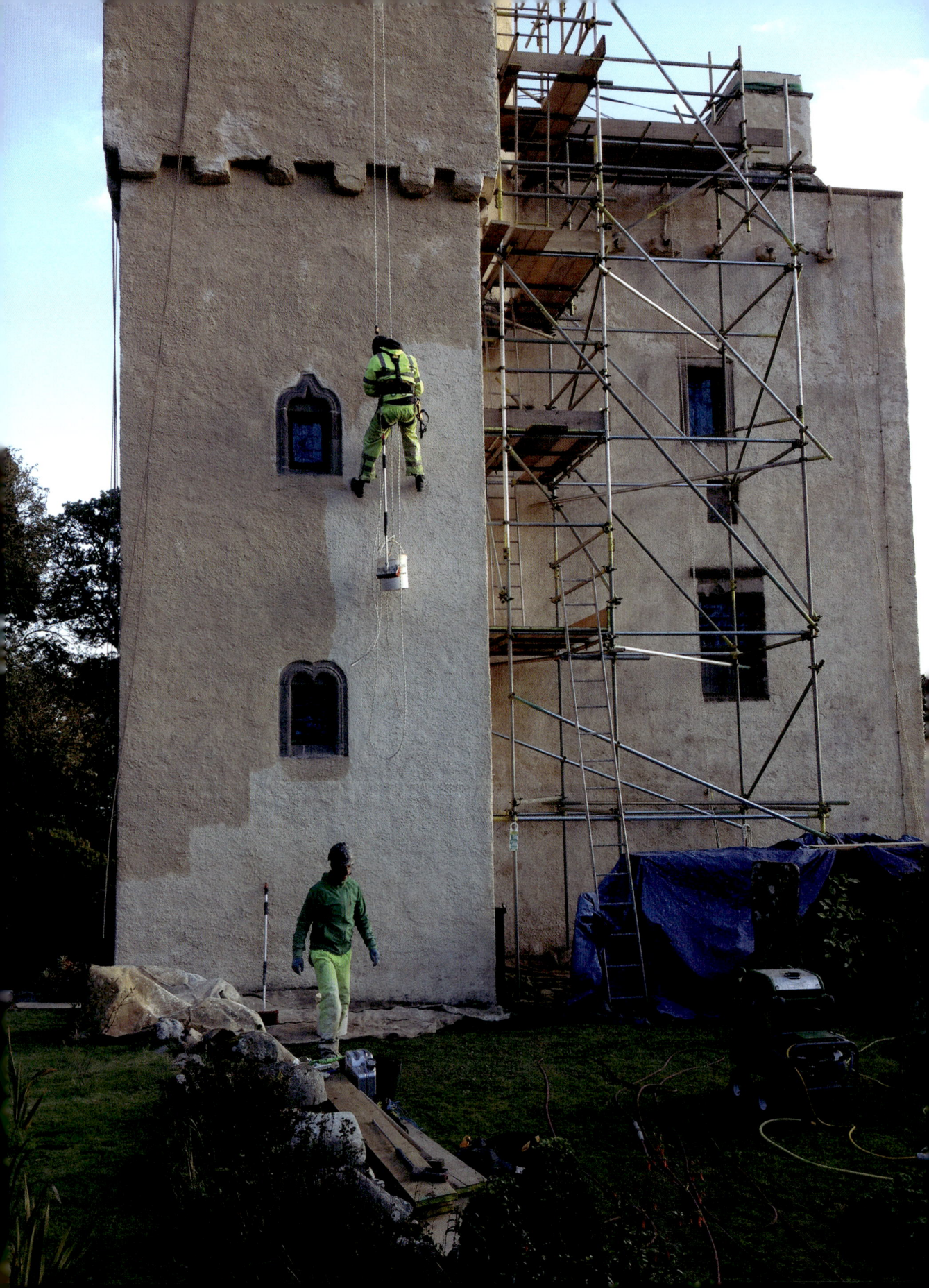

The Exterior

The interior stone walls would have been covered with limewash, a form of paint that is breathable, anti-bacterial and mould resistant. It is made by crushing slaked limestone with various additives. The white walls of Barholm's new interiors probably look as they would have done in the sixteenth century, although we used an expensive modern paint, designed for the interiors of ancient churches, that is both breathable and mould resistant. Outside, however, the walls are painted with genuine limewash every six or seven years. Initially, we hired expensive scaffolding to get the job done, but we have now found Limerich, a company that uses men with rope access qualifications, who mainly work on oil rigs, so the painting is done more cheaply and quickly. It has to be done in the early spring or late autumn, either before the house martins arrive to make their nests in the corbelling, or after they depart again for Africa. We need to book the work more than a year in advance. The exterior always looks bright and beautiful when newly painted, but the paint quite quickly begins to wear away in Scotland's harsh weather. Every five years is ideal for limewashing, but we usually leave it six or seven years, because of the disruption and expense.

Opposite.
Limewashing the exterior by abseiling, 2018.

Harling

Although the ruin of Barholm Castle may have looked 'authentic' in its grey, weather-worn stone, it would have originally been harled and limewashed. There was evidence on the stonework that harling had been used, as was usual practice in Scotland when building with rough stones. Harling is a Scottish word for render, roughcast or stucco. It is a traditional covering for stone buildings, consisting of lime mortar and small stones and seashells, bound with animal hair. The purpose is to keep the rain and damp out, with the breathable harl holding water rather like a sponge in wet weather and evaporating it in the sunshine. Harling is a delicate business, despite it being applied by being hurled onto the wall by hand. It should only be done when the weather is not too wet, not too cold, not too dry, not too hot and not too windy – a tall order in any climate and especially the Scottish one! In practice this means harling in spring or autumn with a good helping of weather-related luck. First of all, the stone walls of Barholm had to be pointed or mortared with a breathable lime mix, since any pointing had been washed away over the past two hundred years. The ruined building was essentially

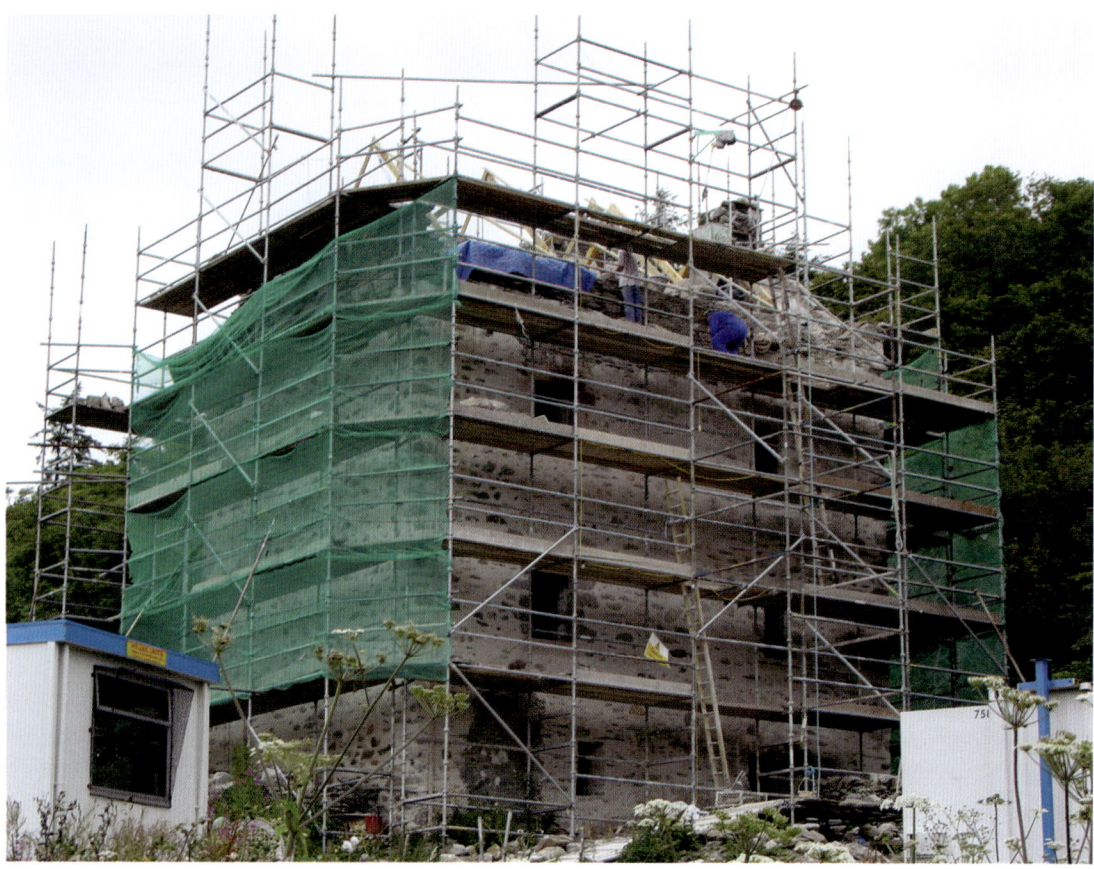

The south side pointed with lime mortar and ready for harling.

four huge drystone dykes joined in a rectangle. Once the pointing was tamped in (packed down) through the 8-foot thickness of the walls, the building was ready for the harling to be applied, at a thickness of no more than 10mm. Netting was draped from the scaffolding to protect the harling from wind and rain – and strong sun, should any happen to break out.

The damage that can be done if the harling mix is not porous enough can be seen in those buildings where cement harling was used for restorations in the 1950s and 1960s and even later, such as Brodie Castle, Culzean Castle, Duart Castle and Kisimul Castle. All of these, and several others, had to be re-harled after the brittle, waterproof cement mortar mix led to water ingress and internal damage, despite the mix having been mandated by the Ministry of Works, later Historic Scotland. It seems incredible that even as late as the 1980s there was such a complete lack of understanding of the basic conservation needs of early buildings and of the use of lime. Earlier, the invention of Portland cement in 1824 led to a craze for its use in house building. This was disastrous for the coating of Charles Rennie Mackintosh's Hill House in Helensburgh, now cared for by the National Trust for Scotland and currently protected by a huge 'box' to allow the precious building to dry out.

The 'authenticity' of grey stones was brought into sharp focus at Stirling Castle, when a bold decision was made to bring its Great Hall back to life, with restoration taking place between 2001 and 2011. The cost was enormous and the outcome magnificent. In its heyday, the exterior walls of Stirling's Great Hall were coated with a colour known as 'King's Gold', an exuberant tone that would have gleamed upon the hilltop from miles around. The glory of the Great Hall began to fade as the paint, battered by weather and time, faded to reveal the grey stonework that can be seen today on most of Scotland's ruined castles. Many Stirling residents expressed dismay at the old grey stones being 'spoiled' by the restoration of the paint. James Crawford addressed this issue very wisely in his essay 'Cool Scotia' on the Great Hall of Stirling Castle:

> When it comes to buildings 'restoration' is a loaded word. It provokes. It asks questions – and often gives answers that we may not like or want to hear. It pulls at the threads of our history, culture, politics, community and identity. It is, in the broadest sense, unsettling. And it is also, fundamentally, about that most personal, elusive – and often exclusive – of concepts: taste. (Crawford 2017, pp. 92–3)

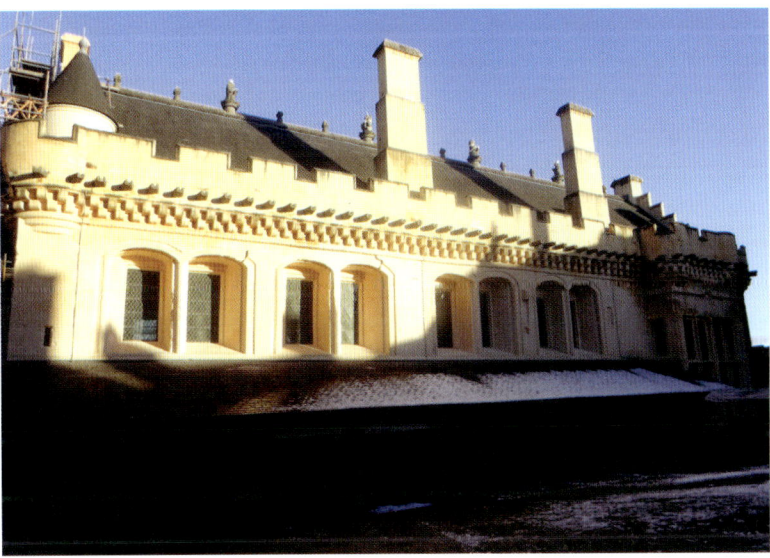

The Great Hall at Stirling Castle, harled and painted in 'King's Gold' after its restoration.

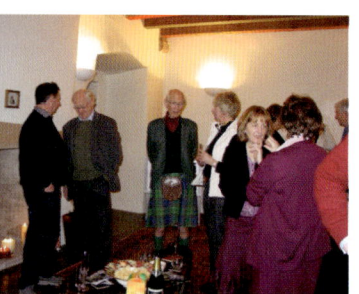

Housewarming party in Barholm's great hall for friends and neighbours, New Year 2006.

I was interested to note that in the Canmore entry for the Isle of Whithorn Castle, there is the claim, written in January 1973, after a visit by the Ordnance Survey, that the harling had been removed, 'thus restoring the castle to its original . . . stonework'.

A New Life for Barholm Castle

Holiday letting

After the restoration was finished, we started to advertise the accommodation with two holiday letting agencies, Celtic Castles and Scotts Castle Holidays. 'Live like a laird in our sixteenth-century tower' was our strapline. We kept prices low, in order to have full occupancy; we had to book our holiday periods along with everyone else. Our paying guests, who came mainly from the north of England, kept the tower heated and aired from 2006 to 2011 and provided us with a small income to offset the costs of housekeeping, fuel, etc. I have written elsewhere about the guests who came to stay in Barholm Castle (Brennan-Inglis 2014, chapter 6). Managing the bookings from overseas while I was occupied with a full-time job was a challenge. There was no interactive online booking system available; everything had to be done by email and telephone, but both agencies were extremely helpful and together we kept Barholm mostly fully booked throughout the year.

Art, literature and architectural vision at Barholm

Scottish castles, with their romantic charm, have provided inspiration for artists for centuries. We commissioned a number of pieces from individual local artists to celebrate Barholm. Pat Ross sketched a watercolour of the south side before the restoration began, and Andrew Briggs made one of his exquisitely detailed pencil drawings of the north-west aspect. Once the walled garden was established, we commissioned a watercolour, also of the north-west aspect, by local artist Andy McKean, who specialises in architectural drawings of the buildings of Galloway. To celebrate our silver wedding anniversary, we commissioned Adam Booth, artist blacksmith, to produce a stunning thistle gate at the entrance. A few years ago, when we received a legacy from a dear departed friend, we bought a set of giant sandstone poppy heads from Max Nowell which he installed on the castle rock on the south

THE RESTORATION OF BARHOLM CASTLE

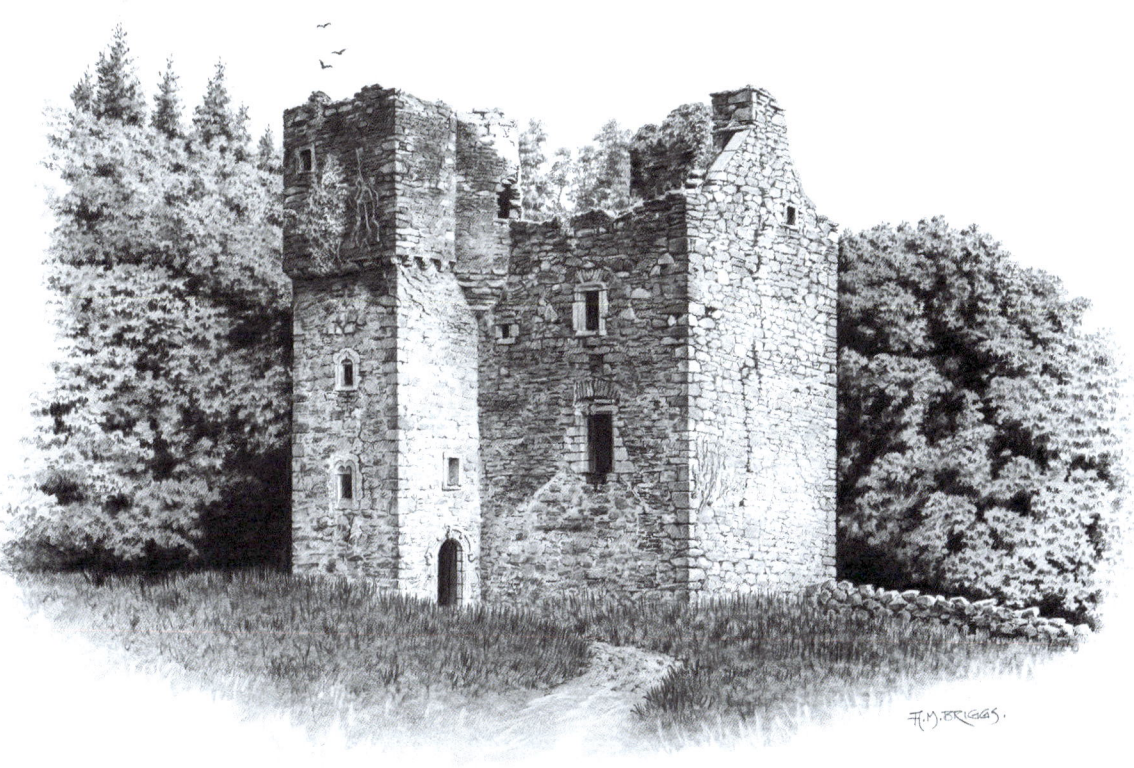

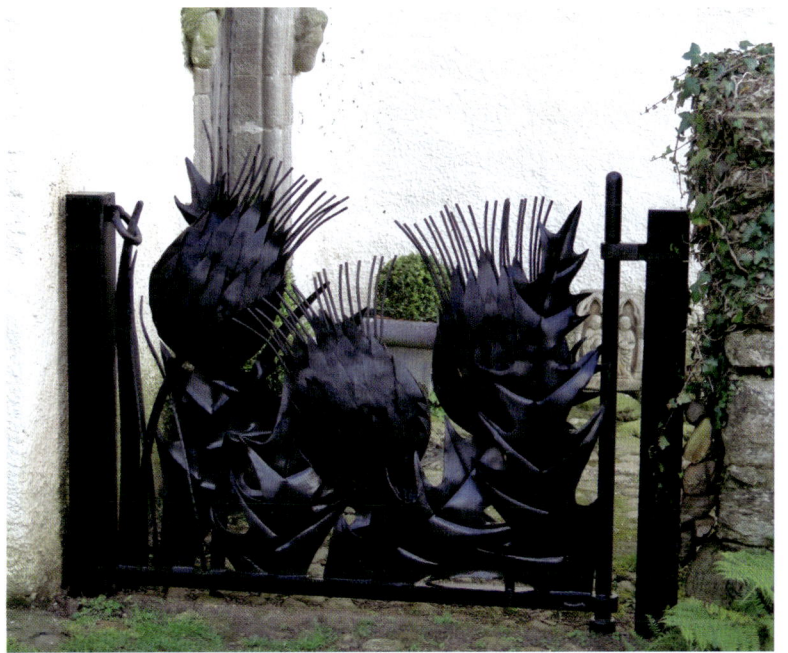

Above.
Barholm sketched by Andrew Briggs just before the restoration started.

Left.
The thistle gate at the entrance to the castle, designed by artist blacksmith Adam Booth.

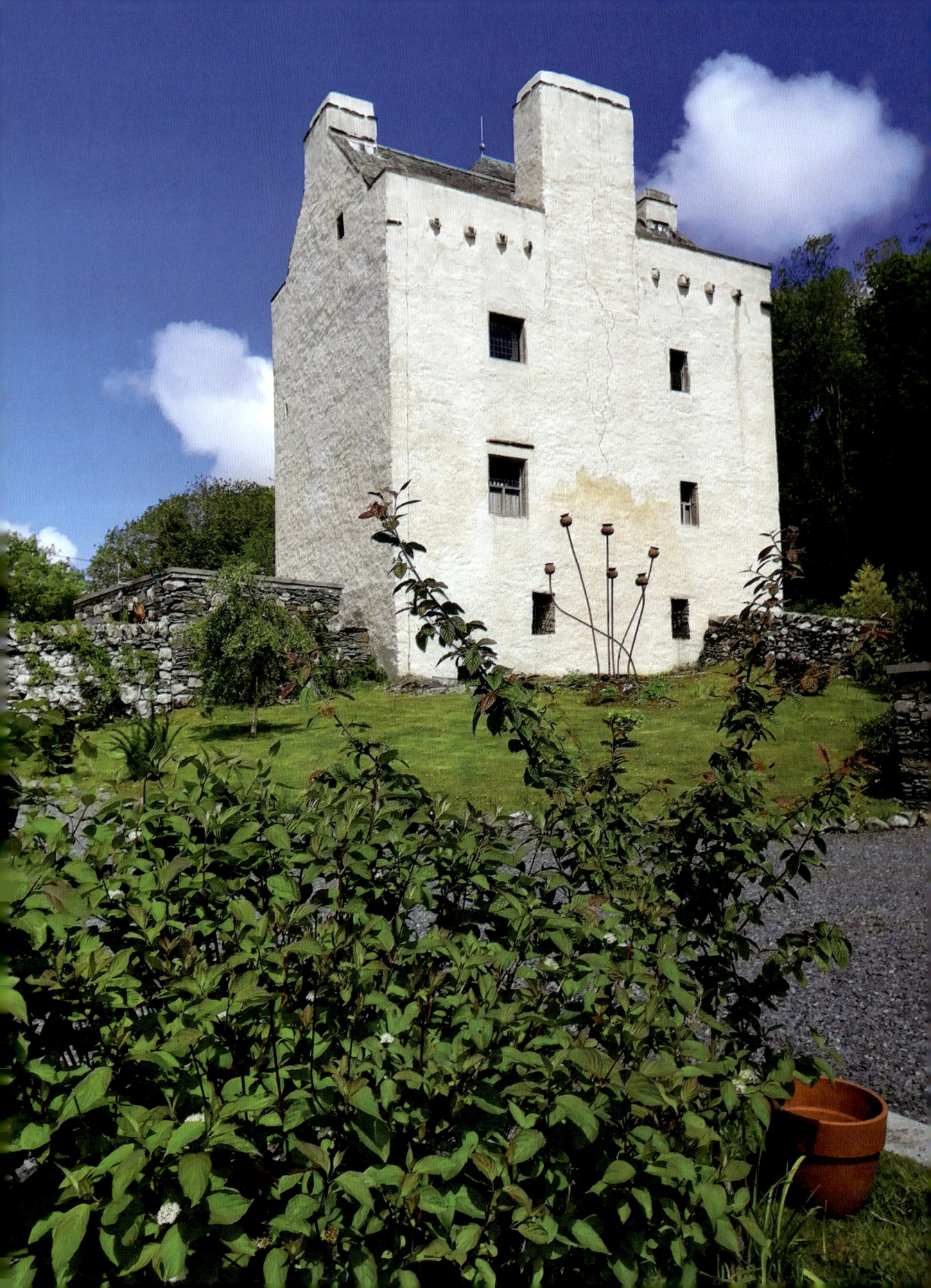

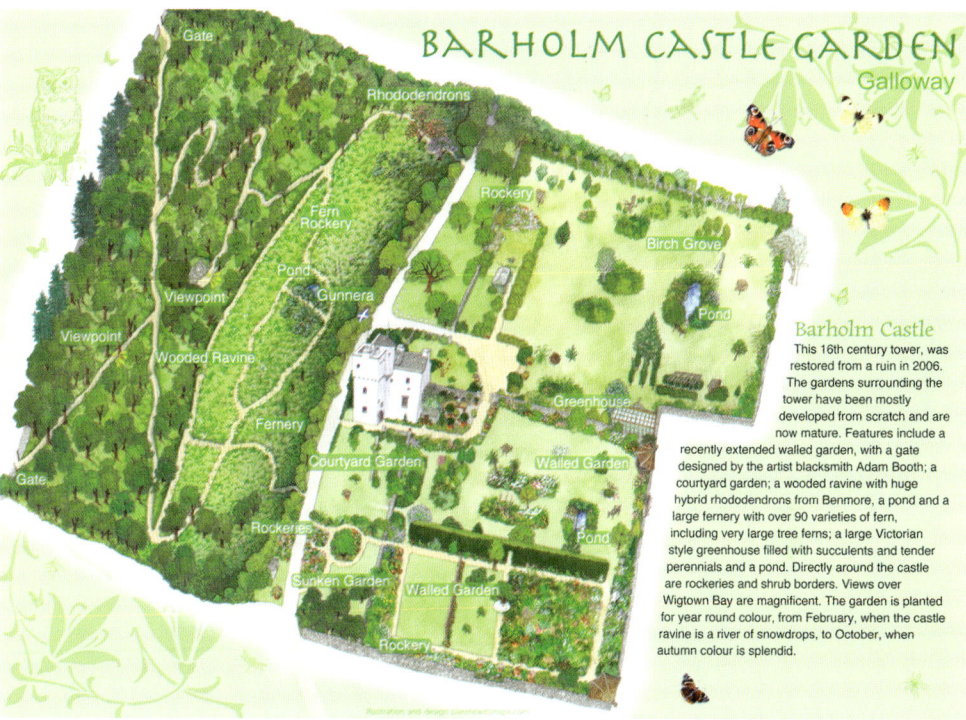

side. In 2022, artist and mapmaker Claire Hewitt came to stay. She surveyed the garden, then painted a beautiful watercolour 'semi-3D' plan of the garden, which we have had printed and give out as a map to garden visitors.

Barholm Castle has been depicted by several local artists – John Copland, Martin Snape and James Faed, among others – in oils, watercolours and pencil sketches; these images have been used as book illustrations and have featured in two well-known novels: *Guy Mannering* by Walter Scott and *Five Red Herrings* by Dorothy L. Sayers.

Several architects sketched out plans of the ruined building before it was finally restored in 2006, and two previous owners had sets of detailed plans drawn up for the restoration of the tower.

One day during Covid restrictions, when the steep woodland paths were still being constructed, I looked up from the track and saw a figure lying on the steep ground halfway up the hillside. My immediate thought was that someone had been exploring from the back road and had fallen and become injured. Anxiously, I shouted up to the person, who replied that she was perfectly OK, thanks, she was just sketching the castle! The artist was Heather Davies, who lived locally; as a result of her visit, she produced a striking watercolour of the east side of Barholm.

Opposite.
Max Nowell's sandstone poppy heads on the castle rock.

Above.
The map of the garden drawn in 2022.

Below.
'Julia Mannering', a character from Sir Walter Scott's novel *Guy Mannering* by Gatehouse artist John Faed (1819–1902).

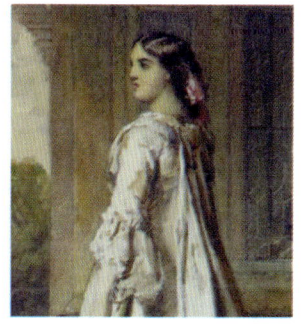

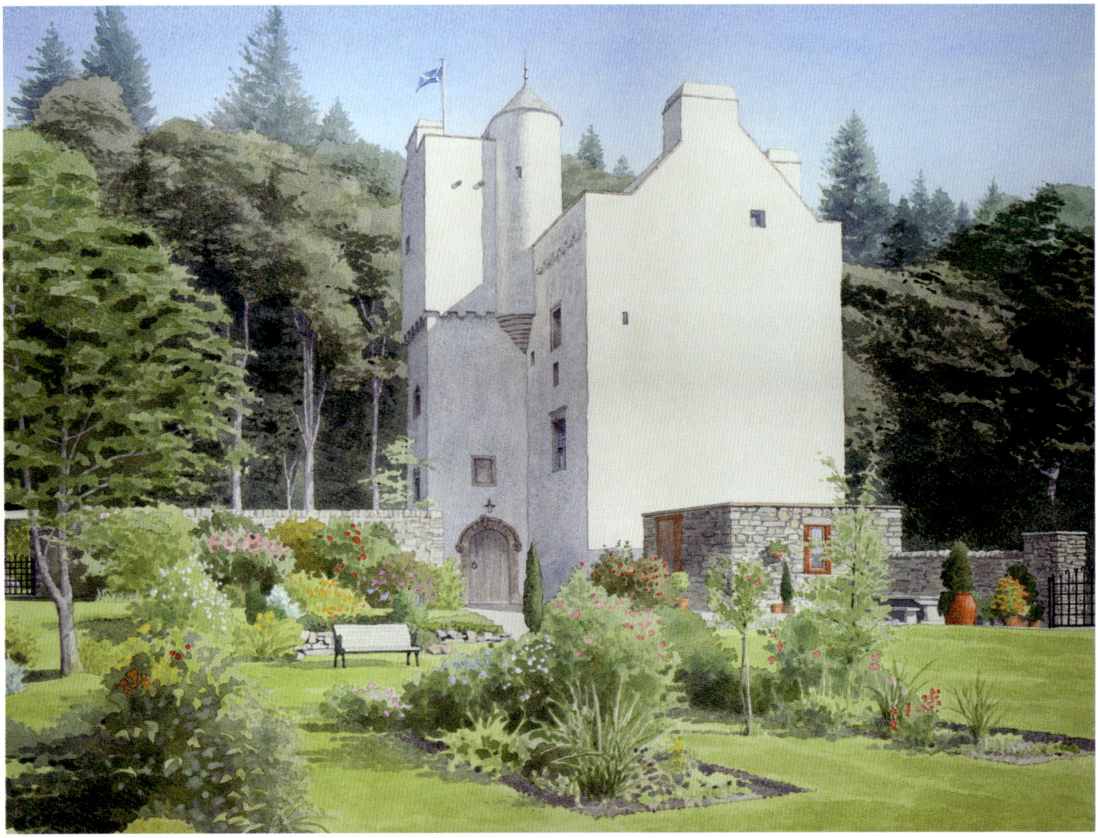

BARHOLM CASTLE

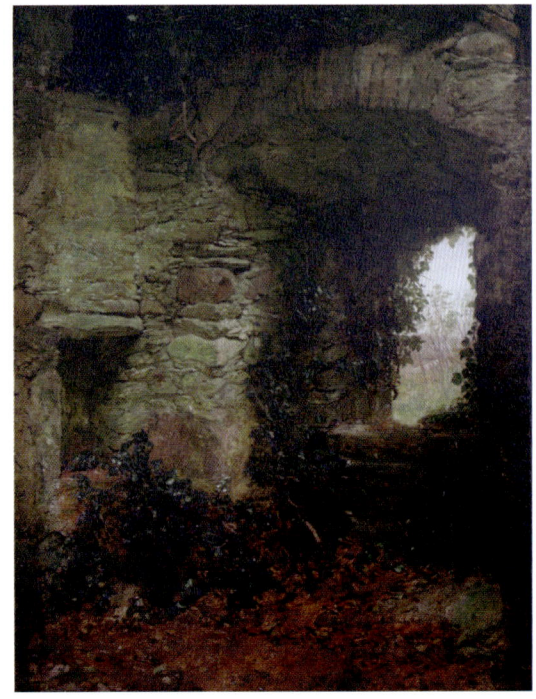

Above.
Watercolour of Barholm Castle and the walled garden in 2014 by Andy McKean.

Right.
The interior of Barholm Castle great hall, south-west corner, before restoration, by Gatehouse artist James Faed, brother of John Faed (1821–1911).

Opposite.
Heather Davies' view of Barholm Castle through the trees, 2021.

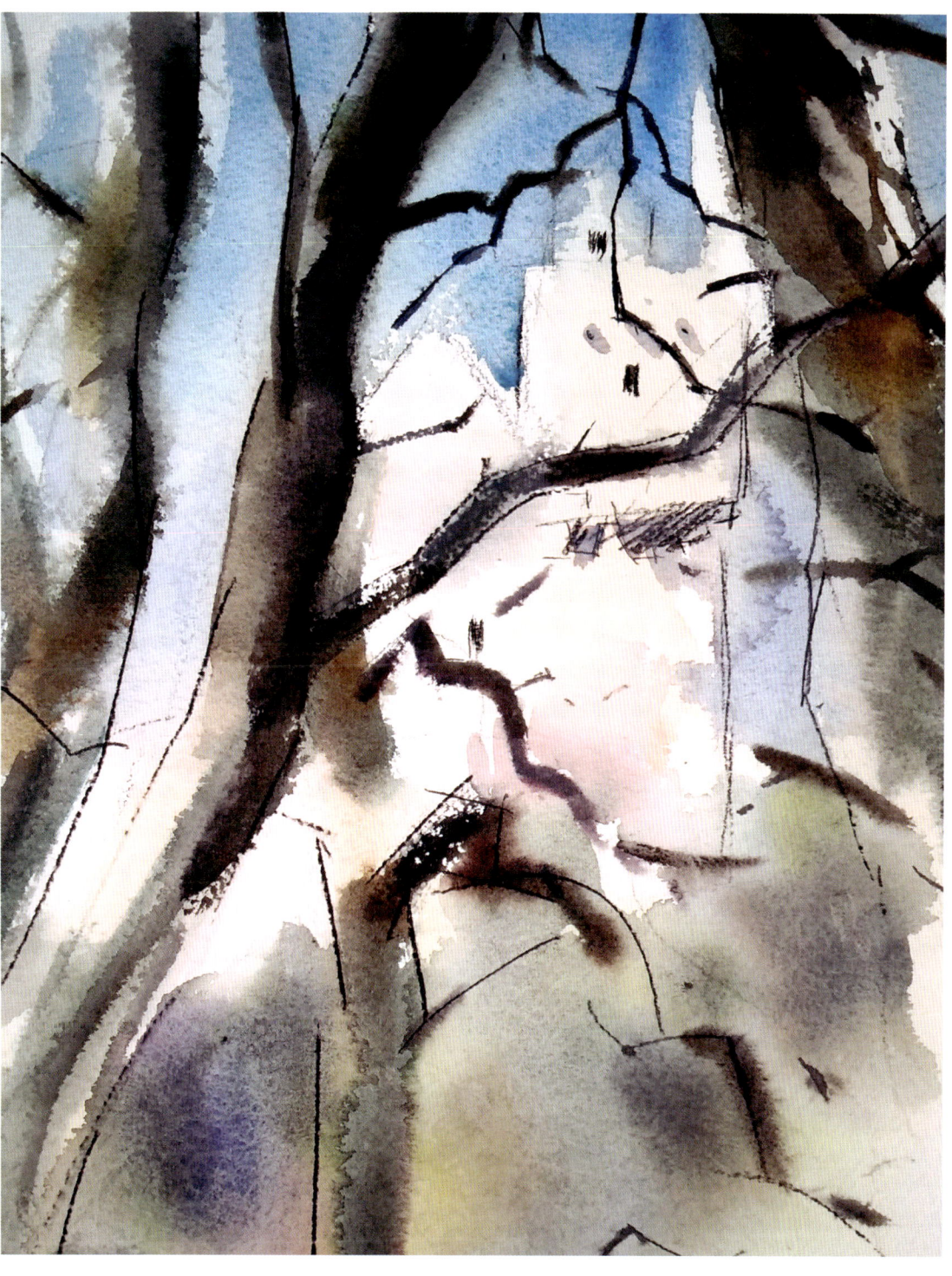

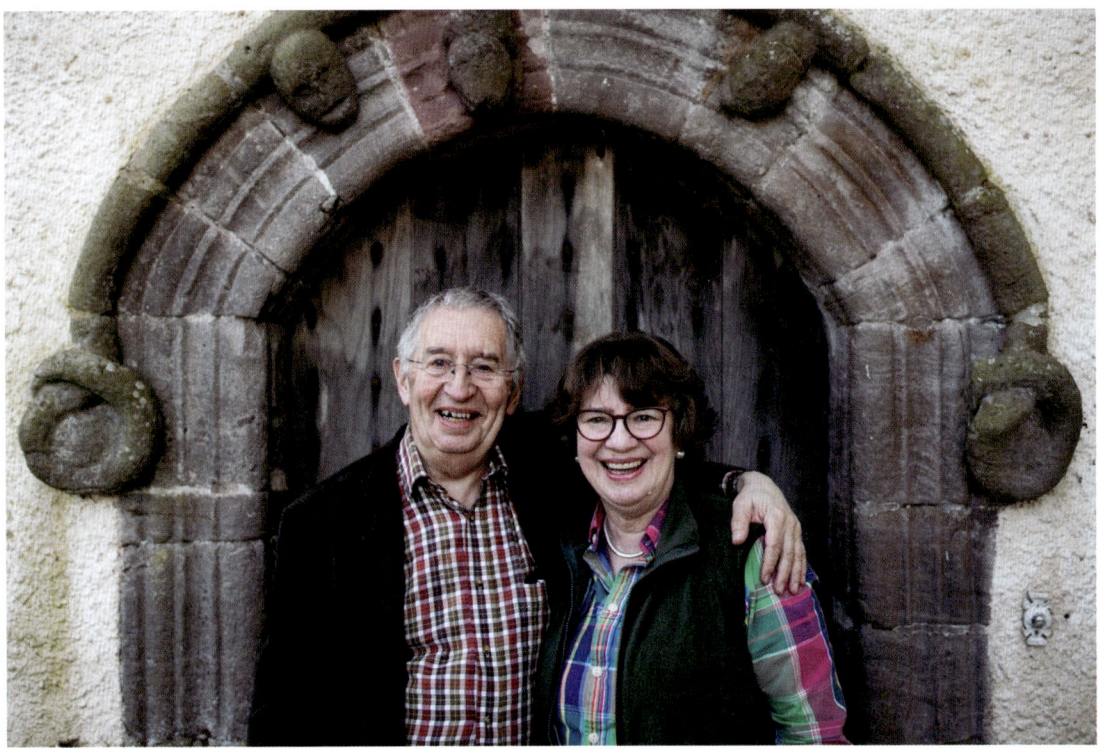

The Brennans at Barholm

The Brennans at Barholm, 2023

We retired from our jobs in the Netherlands in 2011, moved back to Scotland and have lived full-time in Barholm Castle ever since. Originally, the rooms were furnished and laid out to accommodate paying guests, with three double bedrooms and one single (John Knox's room). We have made changes to the internal layout to suit our changing circumstances. First, we divided the main bathroom into two, and created a little office for me in the smaller part. When our daughter Rose graduated from university, I moved my office into her bedroom and moved the washing machine into my old office, which became a laundry room – a great improvement on trekking outside to the cramped and cold boiler house to run the washing machine.

Every year we welcome dozens of visitors to Barholm Castle and to the gardens. We host U3A groups, Rotary clubs, architectural societies and individuals – often McCullochs – who ask if they may visit, and we participate in Doors Open weekend most years. Yes, it's stressful having to rush around and tidy and clean the house before they arrive, particularly since there is nowhere to hide any mess and clutter. However, we find that, without exception, visitors are very appreciative and grateful for the chance to see inside a building that many of them have visited as a ruin or only glimpsed from the A75. On days when I find the stairs a challenge and I'm grumbling that the cobwebs are proliferating and the many leaded window panes need washing yet again, it's uplifting to have an admiring group visit. Someone will always say 'you must be so happy living here!'

We are, indeed, very happy in Barholm Castle and we know how lucky we are to have such a wonderful home. It is a privilege to live in any historic building, although most come with challenges. The stairs are steep, and every journey from one room to another means going up or down at least one flight. I try to be organised, to avoid that sinking feeling when you realise that your book/phone/cardigan is three floors up just when you need it. A tower house is like a bungalow tipped up on its end, but at least we are in no danger of suffering from 'bungalow legs'! The garden also keeps us going. John dotes on his ferns in the ravine and I love having so many different garden 'rooms' to tend. We are fortunate to be in good health and able to stay here as we begin to head into old age – or perhaps we are in good health because of the stairs? Thank you, Barholm Castle!

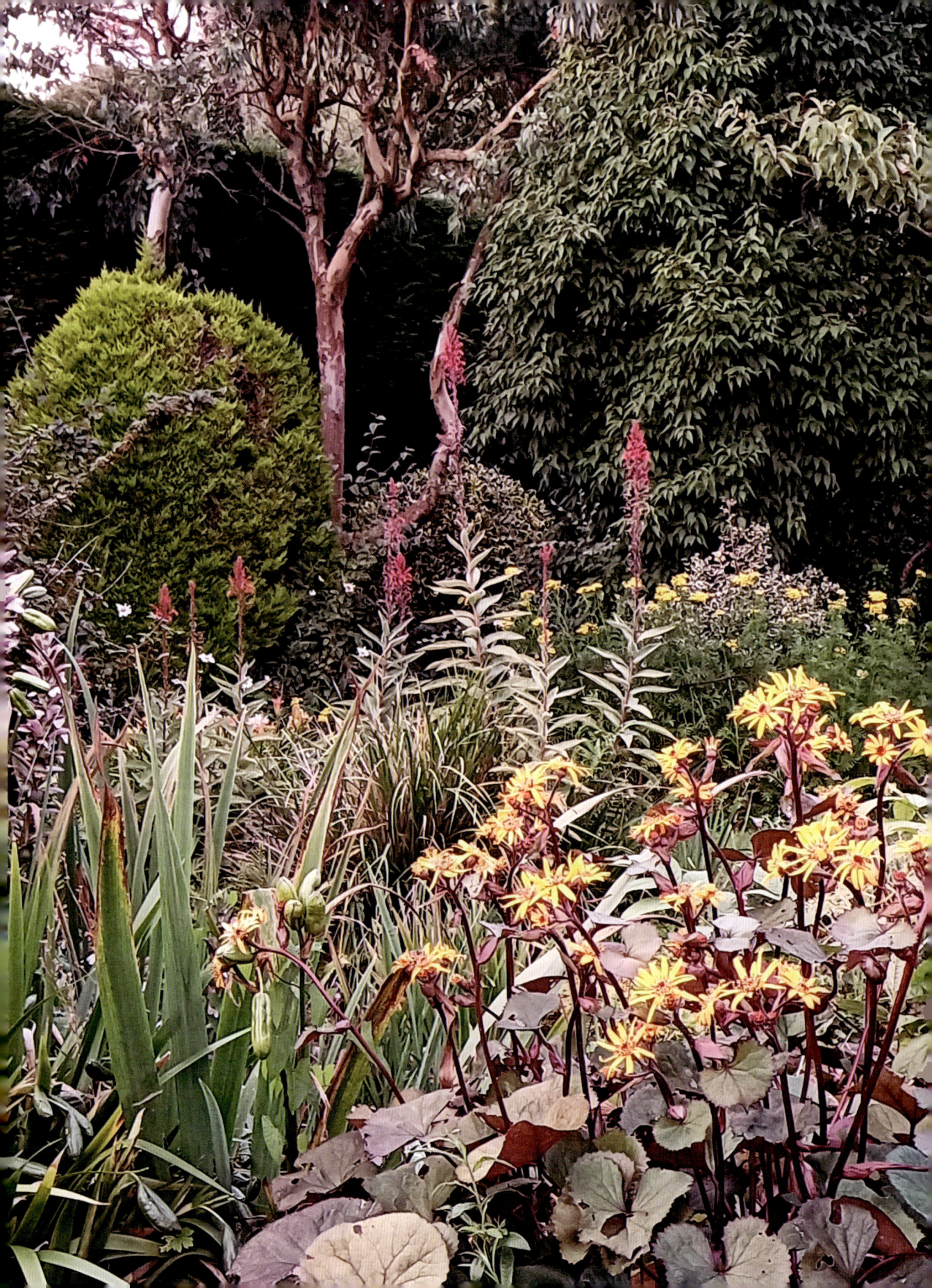

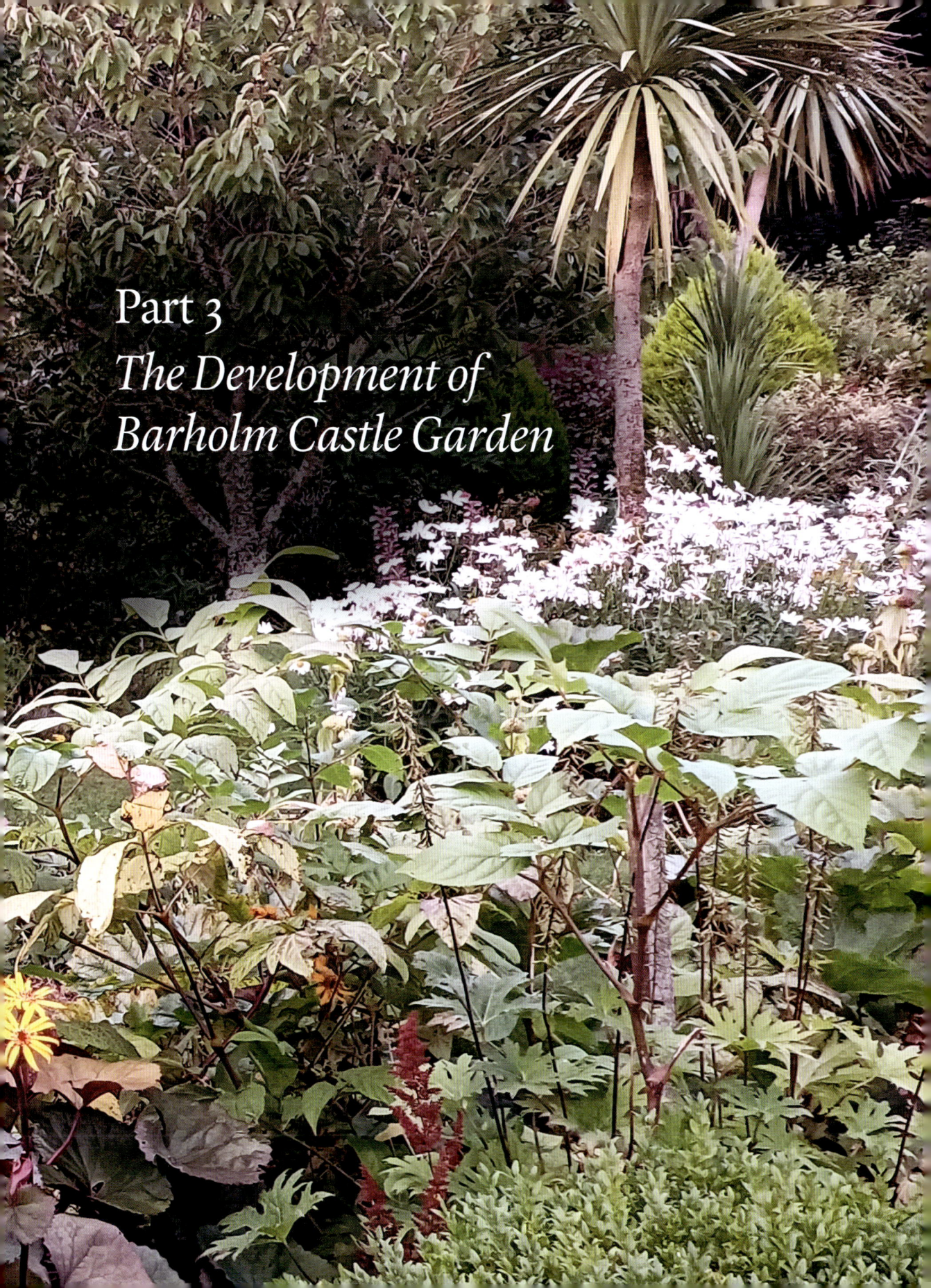

Part 3
The Development of Barholm Castle Garden

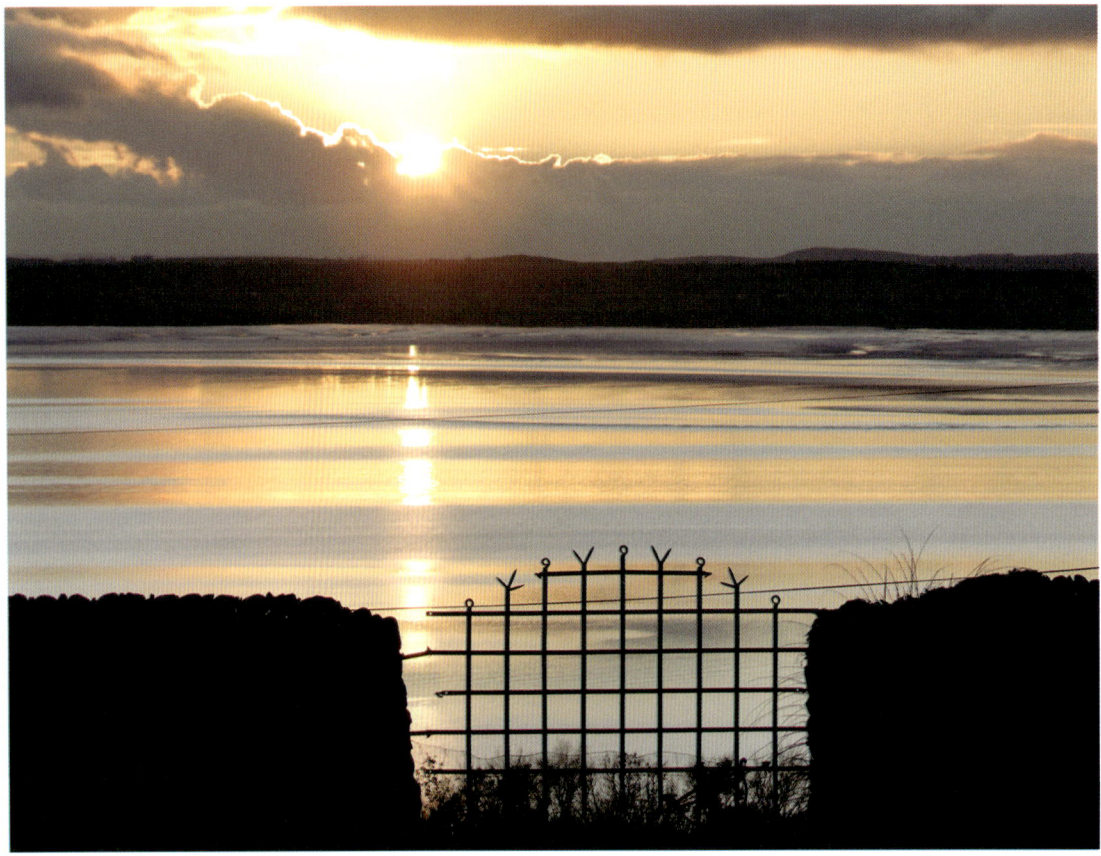

The wonderful view across Wigtown Bay from the walled garden of Barholm Castle; we often have spectacular sunsets.

Introduction

For me, the joy of living at Barholm has always been about the garden around the castle, which we created. During our last year in the Netherlands, I planned the plantings and design from afar, and the first thing I wanted to do on our return trips to Barholm was to check how the garden was doing. Sometimes there were lovely surprises, such as a shrub with unexpectedly colourful growth, or some forgotten bulbs pushing through. More often, though, we came back to an area of planting ravaged by voracious rabbits, swamped by aggressive weeds or wilting from lack of water. On a few occasions we were unfortunate enough to have our stay followed by six weeks without rain, which meant that everything newly planted perished in the drought. Nonetheless, we made good progress from the start, even before the building work was completed.

The garden is in a dramatic setting, almost 400 feet above Wigtown Bay, with spectacular views across Wigtown Bay and south to the Isle of Man. It extends to 3 acres and has six distinct areas. We are very lucky to have a variety of habitat areas. Most have been laid out from scratch by us, with the majority of the planting and garden work having been done since 2005. We have the grass cut for us by a local gardener, Jim Walker, but we do most of the planting, digging and weeding ourselves, which is one reason why it is far from a tidy garden. It is only recently that it seems to have become mature; for years I struggled to encourage weedy little shrubs and trees to grow and despaired when they were cut down by rabbits or deer. It takes at least five years for most large trees and shrubs to establish (buddlejas and eucalyptus excepted), so it was inevitable that I would have to wait for them to make their presence felt.

Roy Strong, creator of The Laskett (Herefordshire), a superb garden, believes that anyone who creates a garden must have a complex series of motives. Initially, having previously gardened only in relatively confined domestic spaces, my aim was simply to grow as many plants as possible in the space available. As an avid collector, I relished the possibility of growing an ever-expanding number of species. For inspiration I looked to Glenwhan Garden, thirty miles west of us, where Tessa Knott has, since 1979, created a fantastic garden out of acres of barren moorland, which is now honoured with an entry in the HES Inventory of Designed Landscapes. I also took inspiration from the tower itself. When we first saw it as a grey ruin it looked impossibly romantic. Now that it is harled and painted, with a Saltire flying from the top of the tower, it exudes a different kind of romance. The castle dominates the walled garden, with the creamy west wall turning a soft pink glow in our spectacular sunsets. An arts-and-crafts aesthetic seems to suit it best. With a crazy paving path to the front, stone terracing and a rectangular pond constructed from local stone, all set off with blousy planting, I think it shows at least a nod to the 1930s designs of Jekyll and Lutyens. We even have a Lutyens bench to sit on and admire the sunsets reflected on the west wall of the tower.

Designing and Creating the Garden

Not much is known about the gardens of small sixteenth-century estates such as Barholm. Did they have pleasure gardens or were the walled enclosures simply there to protect the fruit and vegetable crops from the predations of deer, rabbits and greedy or hostile neighbours? In the sixteenth century,

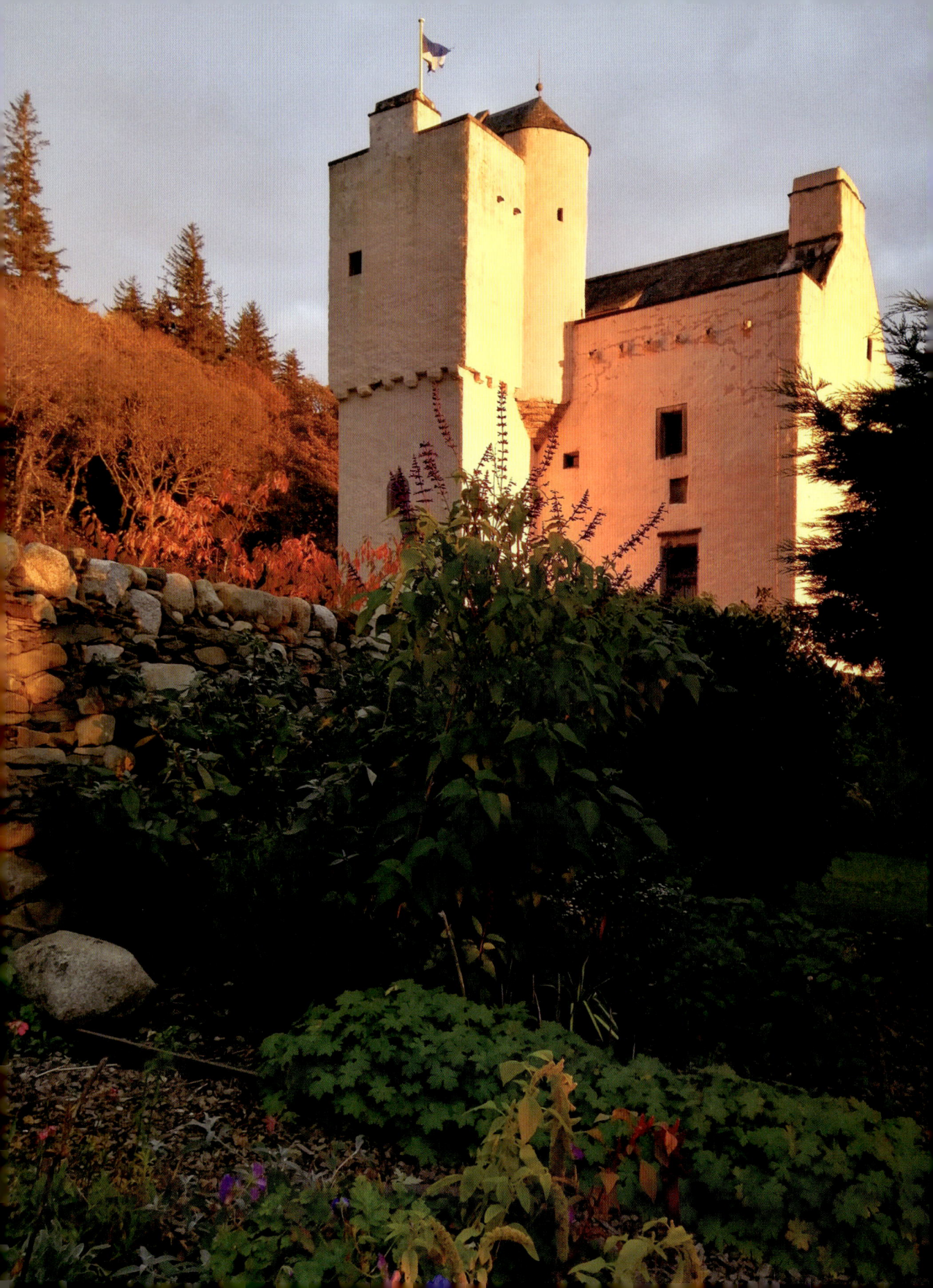

'professional' gardening was carried out by monks, sometimes supported by lay gardeners. Monastery gardens cultivated fruit and vegetables, herbs (both culinary and medicinal) and possibly ornamental flowers; in Galloway, monasteries included Tongland, Whithorn Priory, Dundrennan, Glenluce and Sweetheart Abbey. The knowledge and experience acquired by monastic gardeners would probably have filtered down or been translated across to the non-ecclesiastical estates and houses of Scotland, which employed their own gardening staff.

Some modern owners of sixteenth-century gardens have focused on restoring their gardens to the kind of layout and planting that might have been put in place when the tower was built; this has been the case at Buittle Castle near Dalbeattie, which has a very attractive sunken garden (see page 16). The purist idea of cultivating a garden filled only with flowers and produce available in the sixteenth century is an appealing one. But, in the end, it is not appealing enough to deprive me and Barholm of many of the superb plants that have become available to gardeners over the past 500 years.

Looking through photographs when I put this book together, I saw how much the garden has developed and changed since we started. It is quite startling to see early pictures and remember how proud of a scruffy corner we were at the time, just because we had made such a change from what was there before. It has taken a lot of hard work to shift the garden from all the 'befores' to what it looks like today.

I like the maxim 'live as though you are going to die tomorrow; garden as though you are going to live forever', although it is not quite what we follow. I have my eye, not on the distant future, but on an indeterminate stretch of time about five years hence, when the shrubs and trees I plant will begin to look well established. For several years I literally looked down on most of the newly planted trees in the garden, but now these tower above people enjoying the garden. The garden looks well established, but it is hard to say exactly when the transformation took place from new and struggling garden to one that looks as if it has always been there. Of course the garden changes constantly, as plants die back and new ones take their place. I also know that the garden will not stay as it is when we finally move away. It will become someone else's then and, as the owner of a gardening company, I have seen all too often the outrages some people commit if they do not care about plants and gardens. The mantra of 'low maintenance' has been responsible for the massacre of many mature trees and shrubs, as have the ambitions of up-and-coming designers keen to make their mark on an established garden by transforming it out of all recognition. No matter: we will not look

Opposite.
The walls of the castle glowing in the sunset.

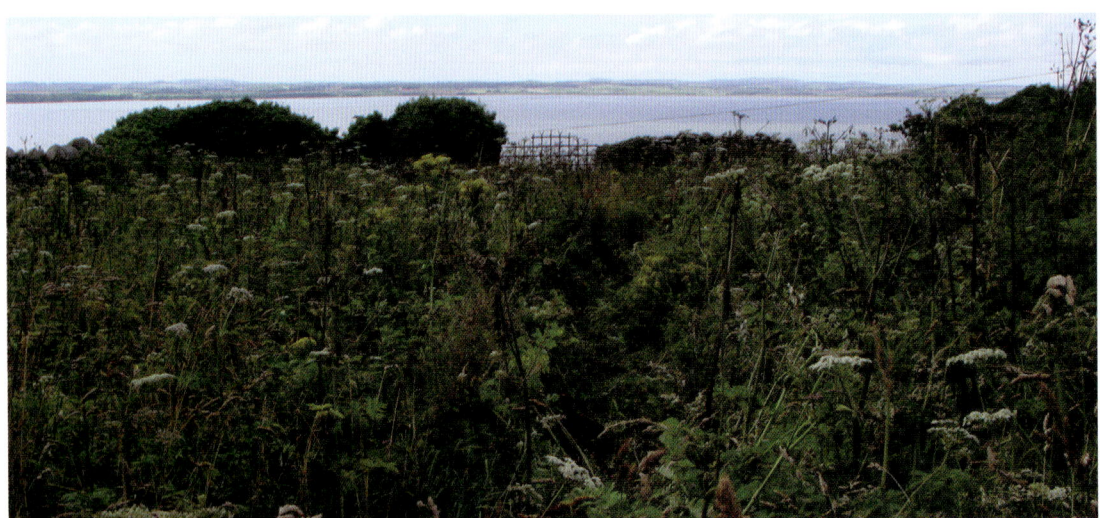

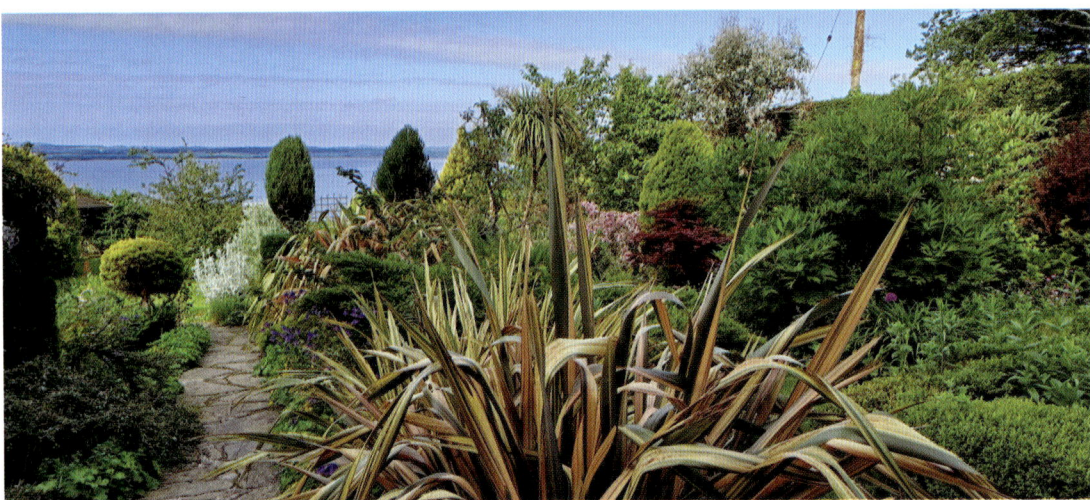

Top.
The walled garden when we first bought Barholm, head-high in thistles and hogweed; the walls were thickly covered in ivy.

Above.
The walled garden now.

back on our ephemeral garden once we have left.

My strategy – if I can dignify my somewhat random approach with such a term – was always to get grass and lawns going first, then work inwards from the walls, planting specimen shrubs and then creating ever-expanding flower and shrub beds along the edges as and when I had the resources. Island beds came later and were always more difficult to establish. Beds mean lots of planting space, but they also mean more maintenance, with weeding a priority until groundcover is established. Beds also need edging, an ongoing job. The grass, and indeed the whole garden, looks much better with crisply defined edges, but keeping them clipped and tidy is a lot of work.

THE DEVELOPMENT OF BARHOLM CASTLE GARDEN

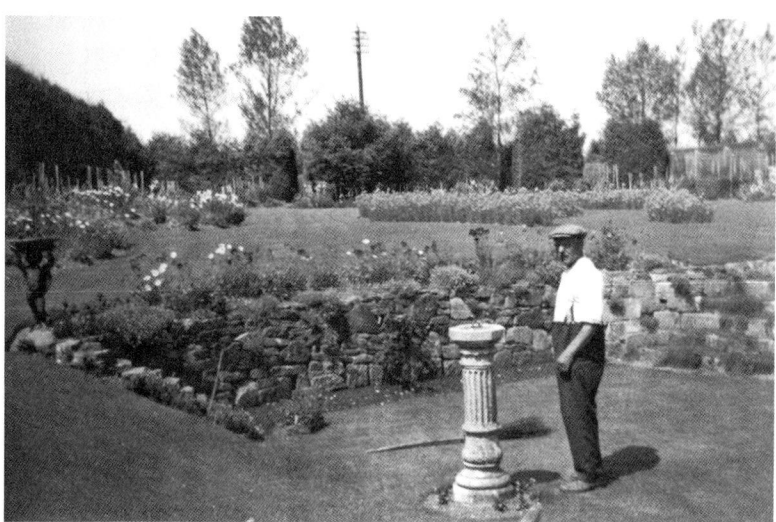

Grandad at work in the 'big house' garden.

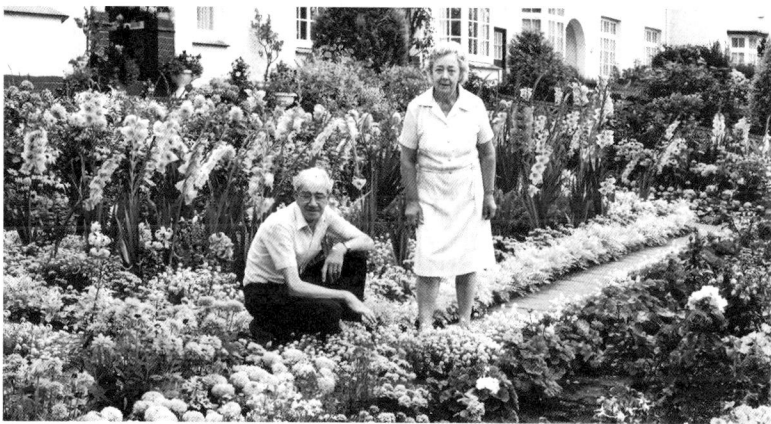

My parents in their front garden in Devon.

A gardening family

I come from a family of gardners. My father used to grow – and show – dahlias, chrysanthemums and delphiniums, gorgeous blowsy blooms with heavy heads that needed staking and support and lots of care and attention. After he retired, he and my mother raised a prodigious number of bedding plants from seed every year and laid out their large front garden in Devon with a riot of colour. One year they were immensely proud to win the 'Torbay in Bloom' competition for their retirement garden. They also grew a great variety of fruit and vegetables; throughout my childhood I ate home-grown fruit and vegetables almost every day.

My maternal grandfather was a gardener, employed by people in 'big houses' to look after their gardens, after he returned from the trenches in the First World War. I remember him fondly and often wonder what he would think of Barholm garden. He and my grandmother treated their long back garden as an allotment, growing all kinds of fruit and vegetables to feed the family. They even kept a goat in the garden and its moth-eaten skin eventually provided a chair cover for Granny.

Planting

A garden is a dynamic thing: it changes not only as the seasons change, but according to the weather, the climate and the effort put in by the gardeners. Plants, like every other living thing, have an allotted lifespan. Each season, some will die while others flourish. Looking back on photographs and lists of plants put in during that first heady period when we started the garden in 2000, relatively few of the early specimens have survived. A combination of rabbits, drought, lack of care (while we lived overseas) and invasive weeds saw off all but the hardiest of plants. Long-distance gardening is not an easy pursuit, and has a low chance of success. However, we would set off several times a year from the Netherlands with the car boot optimistically packed full of plants (those were the pre-Brexit days, when free movement of plants was permitted) and spend our short stays at Barholm weeding and planting, whatever the weather. The development of this garden has needed hard work, enthusiasm and gritted determination in the face of many failures.

Although it was a tiresome business having to wait for five years after the completion of the castle in 2005 to move in, and to be irregular commuters from the Netherlands during all that time, it at least allowed us the luxury of a lengthy introduction to the gardening conditions and possibilities at Barholm. In 2001 we started clearing a tiny patch in the walled garden and introduced a few little plants. By June 2008, I noted in my garden diary that 'our plans are taking shape more clearly'.

I have kept a diary of every plant that I put in. There were a lot of failures, especially in the early years when we lived overseas, but they set the scene and gave us the experience of what grows well and in which locations in our garden. This is a 'no dig' garden: of course I dig holes in order to plant things, but I have never dug-over beds or areas of ground. There is plenty of evidence to show that this is a more sustainable way of gardening, and, besides, digging is very hard work. When we created island beds, our garden helper John Paterson (John P.) hired a turf cutter and then piled up the turves in a corner. The

THE DEVELOPMENT OF BARHOLM CASTLE GARDEN

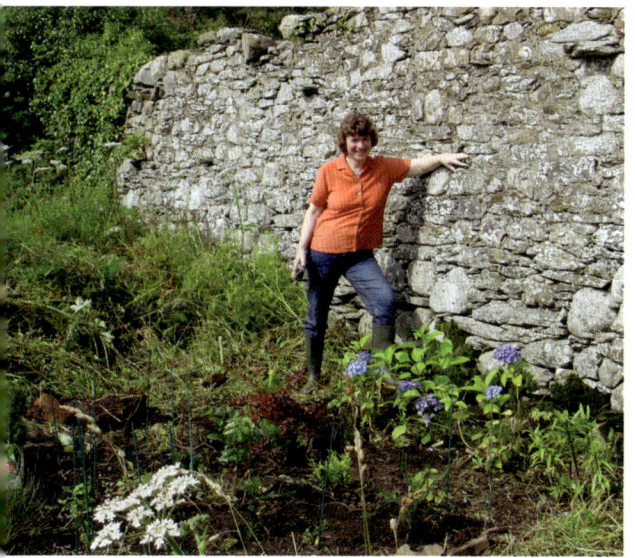 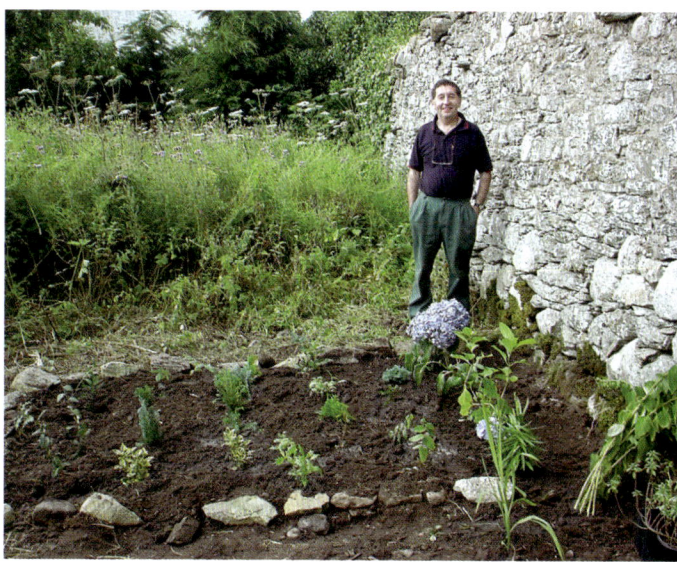

Above.
The Brennans admiring their first gardening efforts in the Barholm Castle garden.

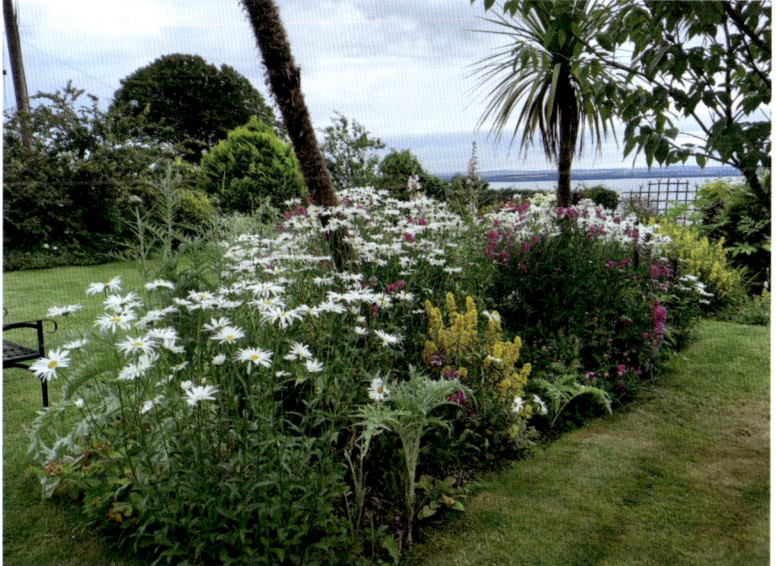

Left.
Establishing new island beds has been very challenging but, when they take off, the results are usually good.

soil there has become rich and loamy so we use it as a resource when we need good compost. I use no pesticides, fungicides, weedkillers or artificial fertilisers. Barholm garden is a pollinators' paradise, with hundreds of different cultivated species and varieties and many self-seeded 'weeds', wildflowers and garden escapes. Biodiversity is a natural by-product of my collector's habit and my reluctance to pull out anything that is decorative and colourful.

If you are planning to open a plant nursery (we considered the idea when we first retired), the last thing you want is a sloping site – your pots will blow over in the wind and roll downhill in the most annoying fashion. However, as ordinary non-commercial gardeners, we are blessed to garden on a series of slopes. We have been able to build terraces and a series of steps linking our 'garden rooms' together, creating all sorts of discrete and differing areas of interest. Altogether we have installed ten sets of steps in the garden. We even have a 'sunken garden' at the side of the road, below the main track level. Not only does the land in front of the tower slope down towards the sea, but directly behind the tower is a sheer drop of about 25 feet, where quarrying took place, down into the area we call the ravine. Jim Walker constructed a flight of steps, in a most challenging area, down into the ravine. A narrow flat area, which might once, long ago, have been a stream bed or pond, is the floor of the ravine, then the land rises again, very steeply, up to the narrow 'back road' which runs parallel to the A75. This wooded slope is so steep, in fact, that we were unable to access it at all for twenty years, until we had a series of zig-zagging 'Himalayan' paths constructed in 2020, during Covid restrictions. We had previously looked up at a woodland that seemed to be dominated by large sycamores. Then, to our delight, we found that the paths led us into a bluebell wood with mixed trees – oak, ash, sycamore,

Left.
Jim Walker making the ravine steps in 2010.

Right.
Barholm's bluebell wood.

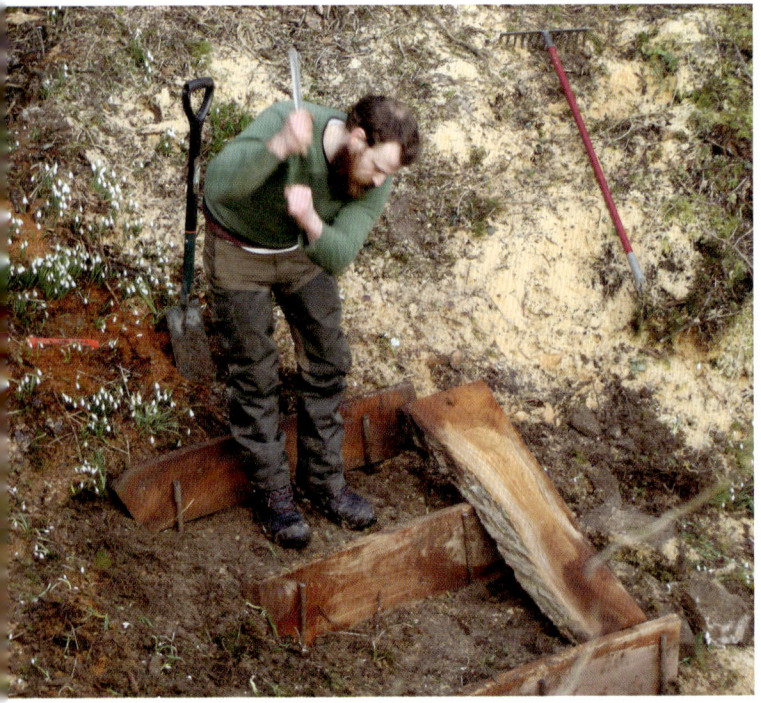

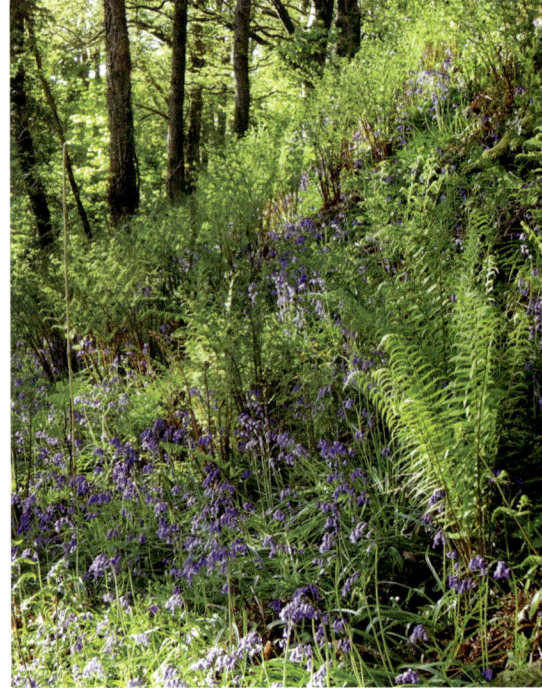

THE DEVELOPMENT OF BARHOLM CASTLE GARDEN

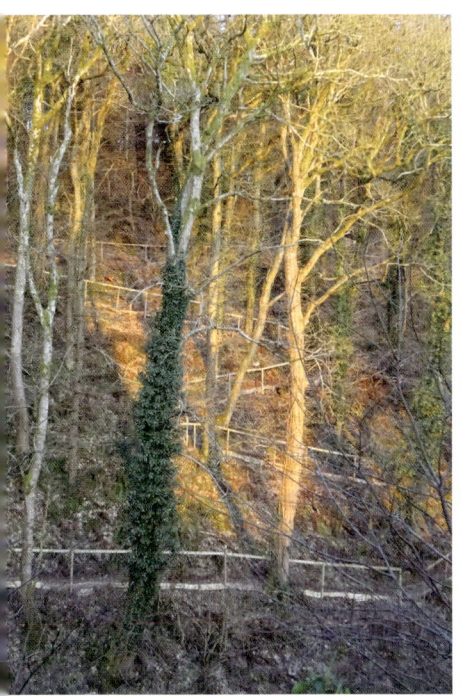
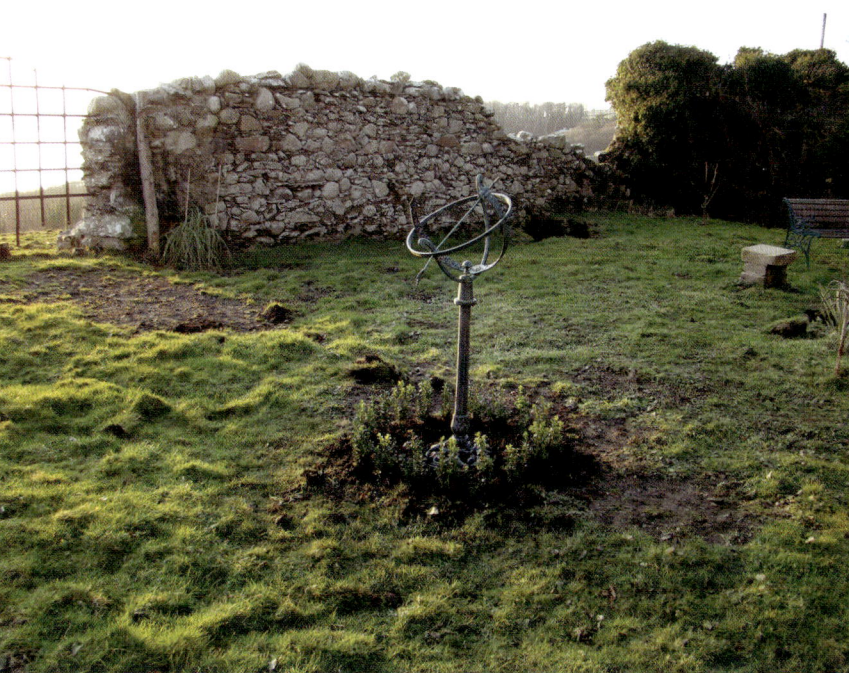

rowan, holly and beech. We now also have a few elms regenerating from those felled twenty years ago when Dutch elm disease struck.

When I first started planting, before we even moved to Barholm, I was grateful for anything robust that would survive on neglect, but gradually I started to acquire a few more special and challenging plants. We are fortunate to have acid soil in most of the garden, although anything acid-loving – such as rhododendrons – planted near the limewashed walls of the tower, or near the parts of the dykes that have mortar in them, will not thrive. Because we are so close to the sea and have the shelter of the walls, and because we have acid soil, we can grow special plants that are borderline hardy or simply will not grow in other parts of Scotland – and mainly they grow very large. *Tetrapanax*, *Grevillea rosmarinifolia*, bottlebrush, figs, cordyline, *Lobelia tupa*, *Eucryphia*, *Crinodendron*, *Fremontodendron*, *Cercidiphyllum* (katsura or toffee apple tree), *Liriodendron*, *Embothrium*, *Hoheria* and *Echium pinana* all grow well in our sheltered coastal site. In the ravine, tree ferns *Dicksonia antarctica* and banana plant *Musa ensete* thrive, although they need some winter protection. In the cold spells of recent winters, gardens in the villages not far east and west of us have suffered casualties, but we have been very lucky and have survived almost unscathed.

Left.
The 'Himalayan' paths through the woods.

Right.
Early days in the walled garden.

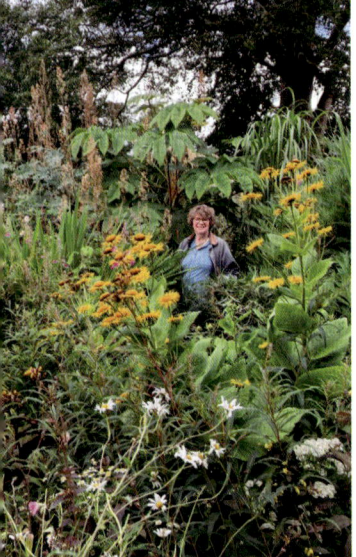

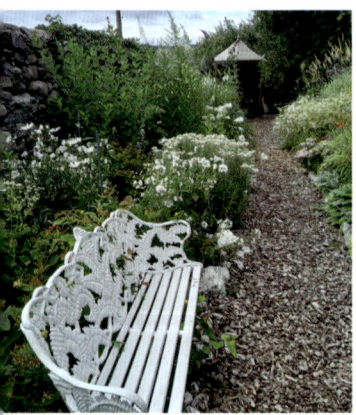

Top.
In the jungly part of the walled garden. This is the large 12m × 12m perennial bed that I started in 2020. Behind me is a *Tetrapanx*, and the yellow daisy-like flowers in front of me are *Telekia*. *Macleaya* and *Miscanthus* also provide height.

Above.
The white border with its fern bench, a reproduction Victorian Coalbrookdale design. A green version of the bench is situated in the ravine, among the ferns.

I love colour in the garden. Nevertheless, when we bought the second part of the walled garden (see page 126) and I had the opportunity to develop large beds from scratch, I decided, for the fun of the challenge, to try a white border against the west wall. It has been a great success, in that a wide variety of white plants thrive there all summer, although I had reckoned without my propensity to allow self-seeders to take up residence. Forget-me-nots and pulmonaria offer blue tones in the spring, and yellow fennel flowers and purple buddleja temporarily strike a jarring note. The white plants include *Fuchsia*, *Persicaria*, *Actea*, giant tree lilies, white rosebay willow herb, *Leucanthemum*, grasses, Japanese anemone and a white-berried *Skimmia*.

But the white border, with its delicate, restrained palette, is not my favourite. I like the drama of deep, rich orange, shocking pink, scarlet and crimson throughout the garden. In early spring I have many planters filled with bright tulips, then, as they go over in late spring and early summer, there are the 'Red Charm' peonies, perennial opium poppies and euphorbia 'fireglow'. As the summer progresses, the crocosmias bring orange and red – 'Lucifer' is by far the most impactful. Two different lychnises, shocking pink *Lychnis coronaria* (rose campion) and bright scarlet *Lychnis chalcedonica* (Maltese cross) make a wonderful, lipstick-bright splash wherever they grow; I plan to expand their territory. Lilies are always good for blocks of bright colour. My favourite day lily, *Hemerocallis* 'autumn red' has large flowers which do not turn to a brown mushy slime after they bloom for their one day – a sad characteristic of several day lilies. The giant tree lilies in the walled garden stand eight foot tall against the shelter of the stone dyke and produce huge heads of flowers. Autumn is the best season for reds and oranges, of course. In the flower borders the orange berries of the arum lily are extraordinary *en masse*, and worth waiting through the scruffy shedding of their leaves in June. The berries of *Euonymous europaeus*, the spindle shrub, are a striking two-tone shocking pink and fluorescent orange and the berries of *Cotoneaster watereri* are so prolific that the whole tree turns bright scarlet.

Cree Landscapes

When we came back from the Netherlands, we were asked to take over Cree Landscapes, the landscape garden company that had been keeping the garden going in our absence and doing much of the structural work. The owner wanted to give up and, as we had just taken early retirement, it seemed like the perfect project for us. I fondly imagined finding commissions from

THE DEVELOPMENT OF BARHOLM CASTLE GARDEN

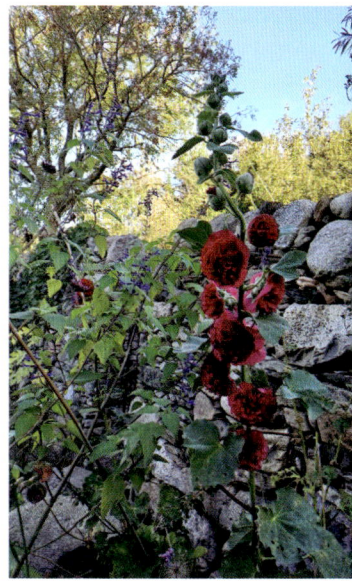

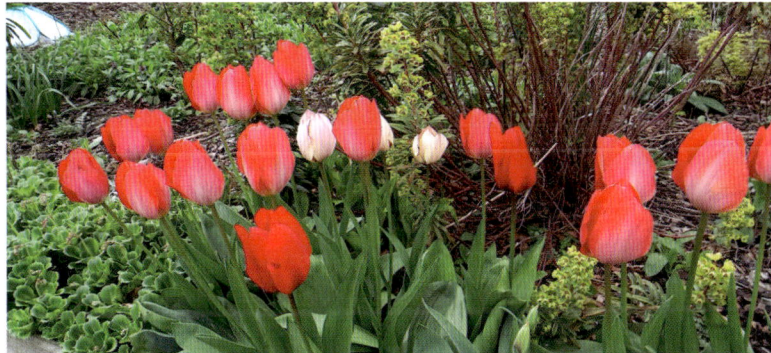

Above left.
Crocosmia 'Lucifer'.

Above right.
Hollyhocks and salvia.

Left.
Tulips and euphorbia.

other keen gardeners in their beautiful gardens, helping to embellish and improve the structure and planting with wise advice and expert plantsmanship, supported by our experienced workers. The reality fell far short of my imaginings, of course. The majority of people who called on our services wanted nothing more than 'low maintenance'. Block paving to replace planting was a constant demand. The really dedicated gardeners of the region do their own planning and planting, and those few new gardeners who want design help tend to source it from Edinburgh- or London-based firms. We lasted for a year before giving up and handing the company over to the foreman, John Paterson, who kept it going very successfully as a one-man company for another ten years before retiring, with Barholm Castle gardens as one of his regular clients. During the time that we ran the company, we

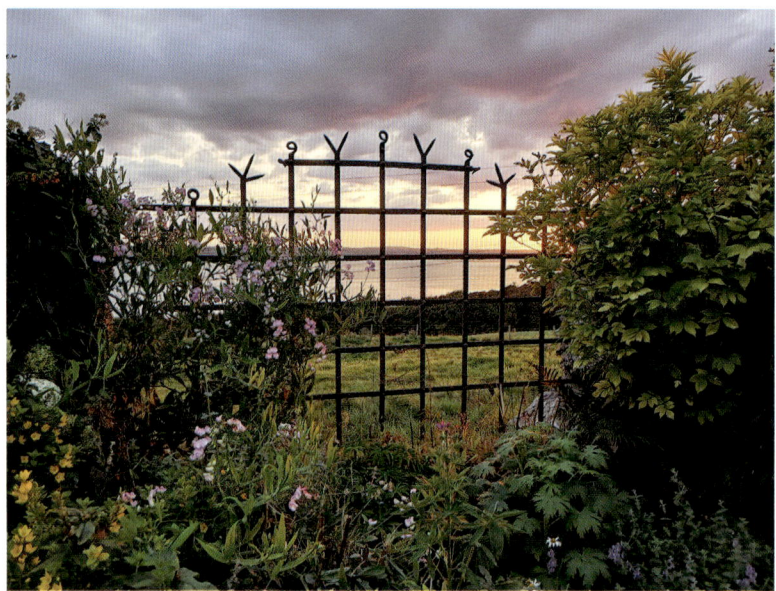

The 'portcullis' at the bottom of the walled garden, actually a plough scarifier, gives us a window to Wigtown Bay.

both learned a great deal about hard landscaping and about estimating the time and effort involved in all kinds of garden maintenance, particularly grass and hedge cutting. This has stood us in good stead in our own garden over the years.

The Areas of the Garden

The walled garden

We are very lucky to have a walled garden – enclosed not by tall traditional eighteenth- or nineteenth-century walls made of brick, but by a six-foot-high drystone enclosure constructed from local field stones, comprising partly the original walls or dykes (probably sixteenth-century), and partly eighteenth-century and later, the lines doubtless dictated by the needs of the farm. Some of the walls have had some mortar tamped in between the stones, and those are the places where acid-loving plants – rhododendrons, camellias, pieris, etc. – turn yellow and fail to thrive. Nevertheless, the walled garden provides us with shelter and the bonus of good soil; the cattle were formerly let in, to graze in what was once a field surrounding the castle, and deposited their manure over the years. Our predecessor had replaced the gate from the farmer's land into the walled garden with a plough scarifier, set immovably into the wall, and it has the look of a mediaeval portcullis. We commissioned the local blacksmith to construct two gates for the walled garden which echo the 'portcullis', and also a matching bench.

The walled garden is in two parts, divided by a high *Cupressus leylandii* hedge. On the south side of the hedge I have planted a variety of shrubs and perennials, which thrive in its lee. This area is laid out in a semi-formal fashion with mixed beds surrounded by lawns, a wooden gazebo and a rectangular pond. There is a low terrace wall with steps in the middle dividing two levels of the garden; this wall is planted with rockery shrubs, perennials and spring bulbs. I have planted lots of large shrubs and small trees in the walled garden, giving structure. The thistle gate into the garden, commissioned from artist blacksmith Adam Booth, makes a spectacular entrance to the castle.

Our gardening had begun in 1999 in the walled garden in front of the castle, where the enclosure would have been an area of shelter for fruit trees and vegetables in the sixteenth century. We were able to purchase only half of the walled garden with the castle, however. The other half had been kept

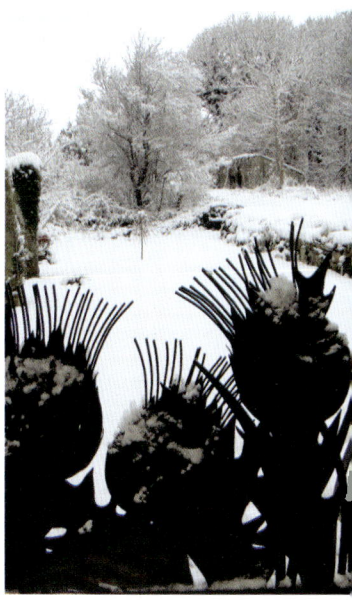

Above.
The thistle gate, commissioned from artist blacksmith Adam Booth.

Left.
Starting to renovate the *Cupressus leylandii* hedge in 2007.

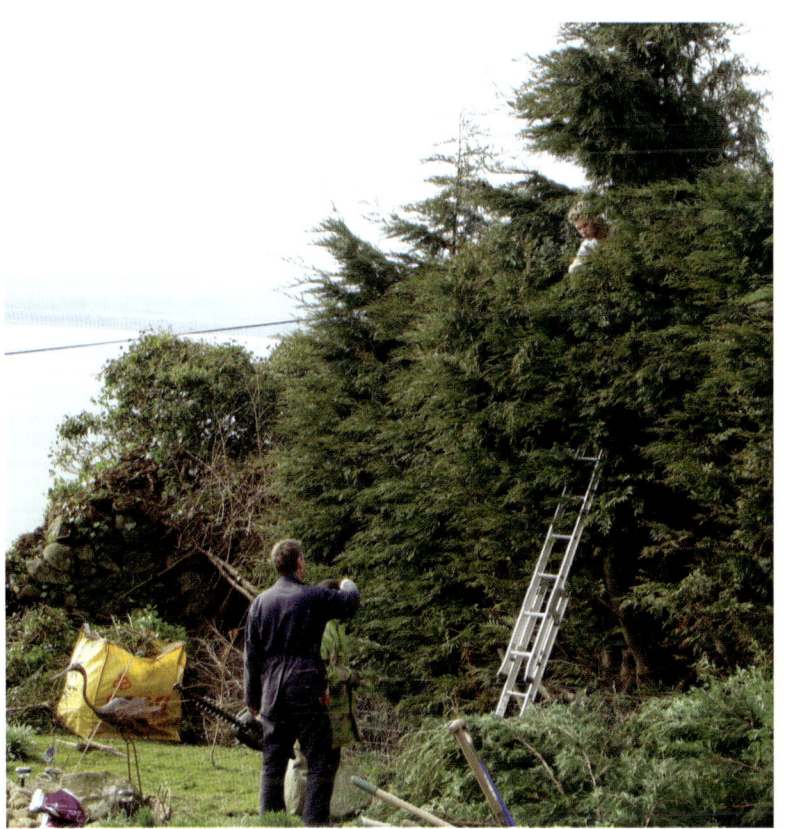

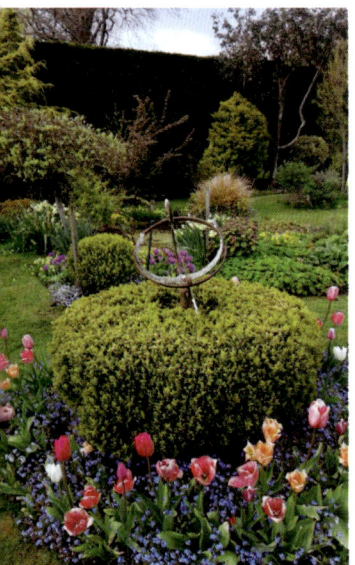

The hedge is a smooth green wall now; it is cut once annually, in October.

back by the farmer in the previous sale, to protect the privacy of the farmhouse, whose garden abutted the walls of the enclosure. The farmer had planted a row of *Cupressus leylandii*, probably in the 1980s, to bisect the walled garden, and this marked our boundary; it was huge and unkempt, until we started to hedge it. Since 2007 we have had the hedge cut annually in late autumn and it now looks glossy and groomed.

The farmers moved out of Barholm Farmhouse in 2015 to their newly restored byre at the end of the track. They rented out the farmhouse and, in order to keep a more manageable plot for their tenants, they fenced off their half of the walled garden. I used to peer through gaps in the leylandii hedge and fantasise about expanding into the unused and increasingly wild space. Happily, when we approached the farmers offering to buy that piece of land, we received a positive answer, and in February 2019 we finally took possession of the remainder of the castle's walled enclosure. Although we had tamed the leylandii hedge on our side, the other side was a ghastly disaster of wayward and dead branches. Being north facing, it had not even had the advantage of sunshine to keep it growing well. When we showed off our garden acquisition and visitors looked askance, assuming that we would be taking down the leylandii, I understood their view. But we were determined to keep the hedge, not only as a marker between the two very different halves of the walled garden, but also as a deep haven for wildlife and as shelter from the winds. The width and height of the hedge is about 9 feet by 15 feet and it houses many birds and beasts all year round. Despite the doom-laden warnings one reads about the difficulty of growing anything in the vicinity of a large old hedge, the ground on both sides – north- and south-facing – has proved to be very fertile and hosts many welcome self-seeders as well as plants positioned there. Our wet climate is doubtless helpful.

The 'new bit', as we still unimaginatively call it, was quite a challenge. It had been colonised by rampant brambles and nettles, all of which had to be removed, along with many pieces of broken glass from an old cold frame, and lots of plastic rubbish. The only plants left were four ancient apple trees and a couple of wild plums, all of which are heavy croppers. I spent the first two years pulling out wayward brambles almost daily, but gradually they diminished to once-a-week jobs and now there is only the occasional one trying to make its way back in. In terms of design, I decided to make a large perennial bed of 140 square metres, bounded by woodchip paths. In it I planted densely. We had spent twenty years slowly nurturing the other half of the walled garden and the paddock, but now we could no longer afford to proceed in such a leisurely fashion. If we wanted to see this new garden flour-

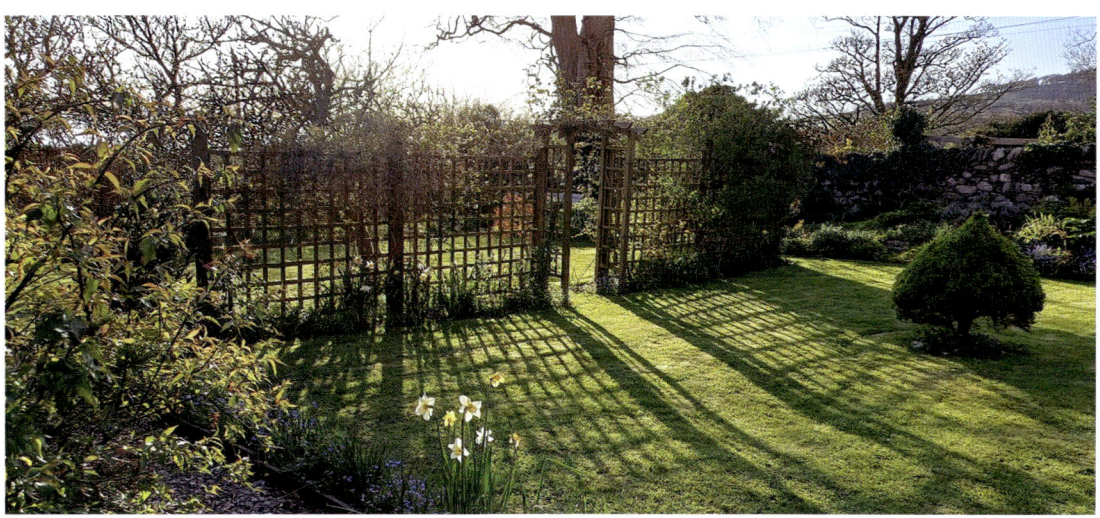

Above.
The trellis and one of the lawns that I sowed. The *Picea conica* in the middle of the lawn came as a gift from the owners of Buittle Castle near Dalbeattie.

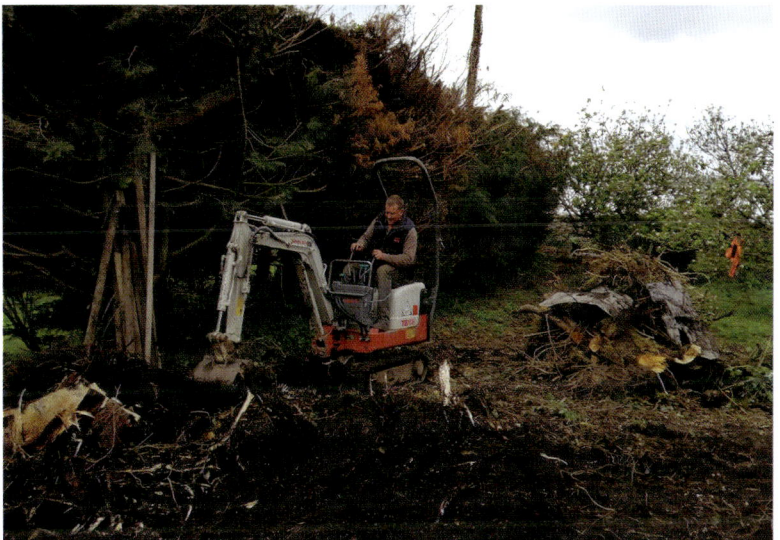

Left.
An unsuccessful attempt to dig up a semi-fallen *Cupressus* that had rooted itself back into the ground. The root and part of the trunk are still there but have now blended into the planting. The north side of the *Cupressus leylandii* hedge is looking at its worst in this photograph. It is understandable why visitors were horrified that we were planning to keep it.

ish, we felt that we had to plant it up quickly. We might not have twenty more years to spare. In fact, within a year the bed was packed with colour, and soon some plants were growing almost out of control, such was the advantage of the good soil and sheltered position. I sowed two lawns, and had a trellis erected between them. It is now smothered on both sides by a profusion of climbing roses, clematis, wisteria, honeysuckle, Virginia creeper and solanum. The soil is wonderful – rich, deep and loamy – because the cattle used to be let in to graze in the walled garden before it was ever cultivated; all those years of fresh manure have paid off.

The paddock

The paddock, a former field, is the largest single area of the garden and has been the most challenging to develop. It has low stone dykes – much lower than the walled garden – and is extremely windswept, but we decided not to plant a shelter belt, unwilling to sacrifice the spectacular views across Wigtown Bay to the Isle of Whithorn and the Isle of Man. Gradually, the large filtering plants that we put in ten years ago along the west side – the pampas grasses, birches, rowans and wild cherries – have made a difference and the paddock has become calmer and less prone to wind damage, even although we are now battered by more fierce storms than we used to experience. The paddock has very poor stony soil, is brutally exposed to the relentless prevailing winds and is home to large gangs of voracious rabbits. They loll around in the sunshine and just glance at me disdainfully as I work to try to limit their damage. At first we planted around the pond, with rabbit-proof fencing to protect the plants. But gardening within a fenced-off bed is tiresome, and the fencing does not look attractive. I now use trunk protectors on trees and shrubs and leave newly planted perennials to take their chances against the rabbits – who often win.

The paddock in 2009.

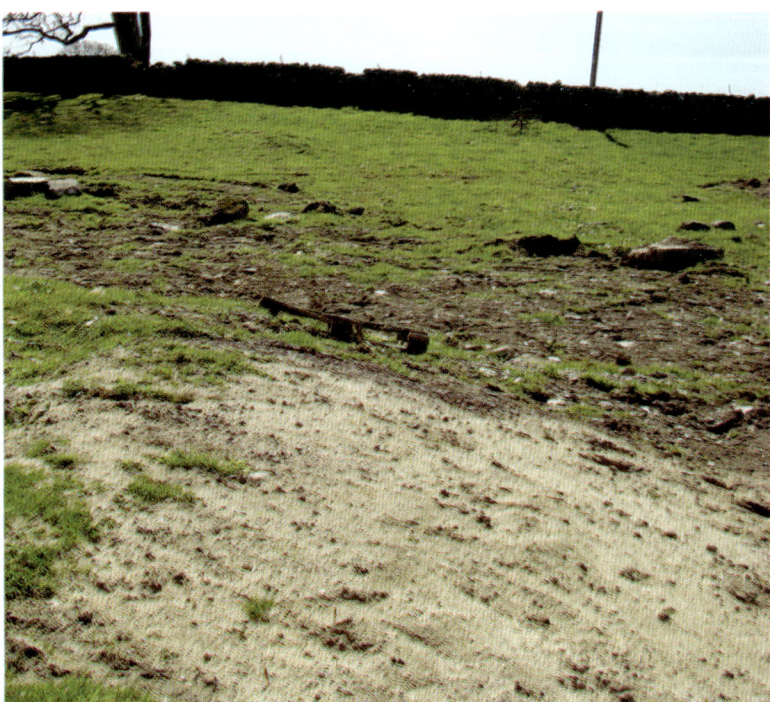

THE DEVELOPMENT OF BARHOLM CASTLE GARDEN

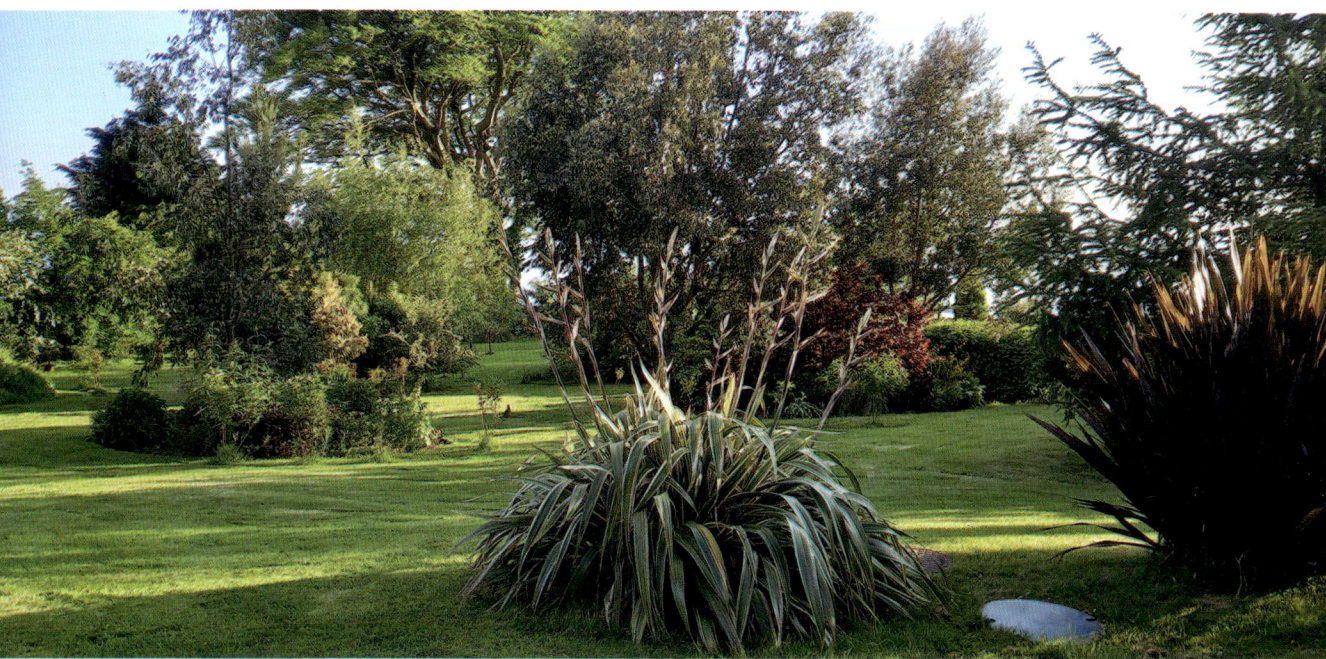

The ravine

At Barholm, we have 'his' and 'her' gardens. I take care of everything above the ravine – the walled garden, the sunken garden, the shrubbery and the paddock – and my husband John looks after the other half of the garden, which is the deep ravine behind the east side of the castle. When we first acquired the ravine, bought in 2004, five years after the purchase of the castle, it was impenetrable, and full of huge standing dead elms which had been stricken by Dutch elm disease. Jim Walker chopped them down and constructed a series of bonfires that burned for days. Between 2004 and 2007 John spent much of the summers dressed in a combat jacket and two thick pairs of trousers, pulling up thousands of 6-foot nettles with his hands encased in fireman's gauntlets. This was the only way to get rid of them; even if we had wanted to use weedkiller, it would have damaged the groundcover of ferns and dog's mercury. The nettles have been replaced with a growing collection of native and cultivated ferns – over one hundred species to date, from all over the world – and large-leaved plants which suit the damp conditions. There are still plenty of nettles left for wildlife on the periphery and in the thousands of acres of land around Barholm.

The paddock in 2023.

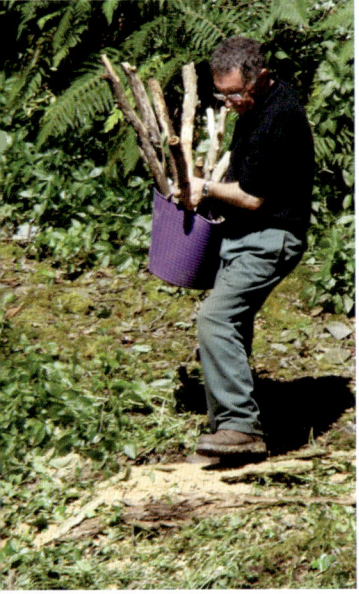

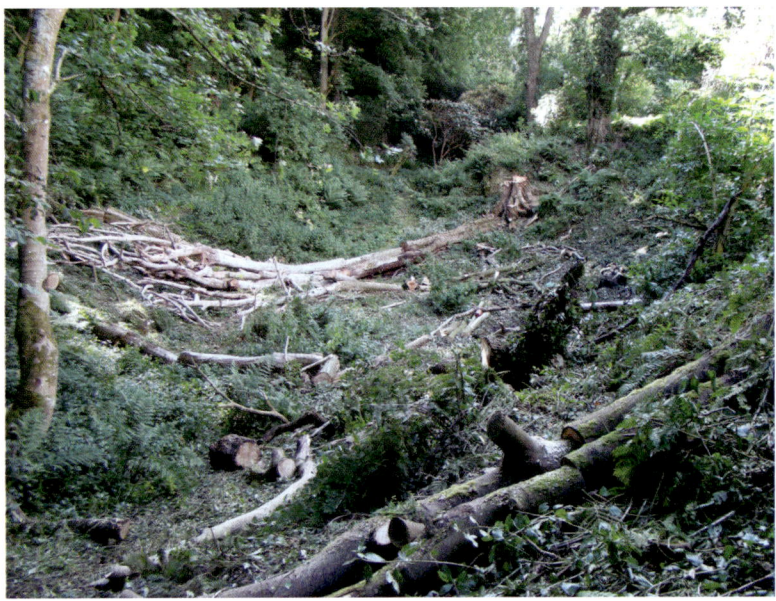

Above.
John making clearances in the ravine.

Above right.
Felled elms in the ravine, stricken with Dutch elm disease.

The ravine, a deep bowl within the garden, is completely sheltered from the winds that often batter the rest of the garden, and the soil is a rich, deep, acid humous, a perfect growing environment for ferns and other moisture-loving plants. The native ferns already present include the male fern, the golden male, the lady fern, the tassel fern, the hard shield fern and hart's tongue. Some of the more spectacular and special cultivated ferns include the giant tree ferns, *Dicksonia Antartica*, the tatting fern, *Athyrium felix-femina* and the shuttlecock fern, *Matteuccia struthiopteris*. In addition to the ferns, there is a large gunnera bog, along with rheums, *Darmera pelltata*, yellow skunk cabbage (bought before it was classified as an invasive alien species), macleaya, huge hostas, veratrum, filipendula, *Podophyllum versipelle* 'spotty dotty' and bananas. Trees we have planted include willow, acer, amelanchior, magnolia and liquidamber and shrubs are mahonia, deciduous azaleas and sweet-smelling skimmia. At the end of the ravine near Barholm House are the huge Benmore hybrid rhododendrons planted by Mrs Hogg in the 1970s. They make a spectacular show of colour from April through to June. There are also a few other mature mixed shrubs, including pink lilac, *Pieris*, *Camellia*, *Philadelphus*, *Eucryphia*, *Crinodendron* And *Mahonia*.

In early spring, most of the vegetation dies away and the ravine hosts a carpet of snowdrops, which are later supplanted by wild garlic and dog's mercury.

We do help each other from time to time in our separate halves of the

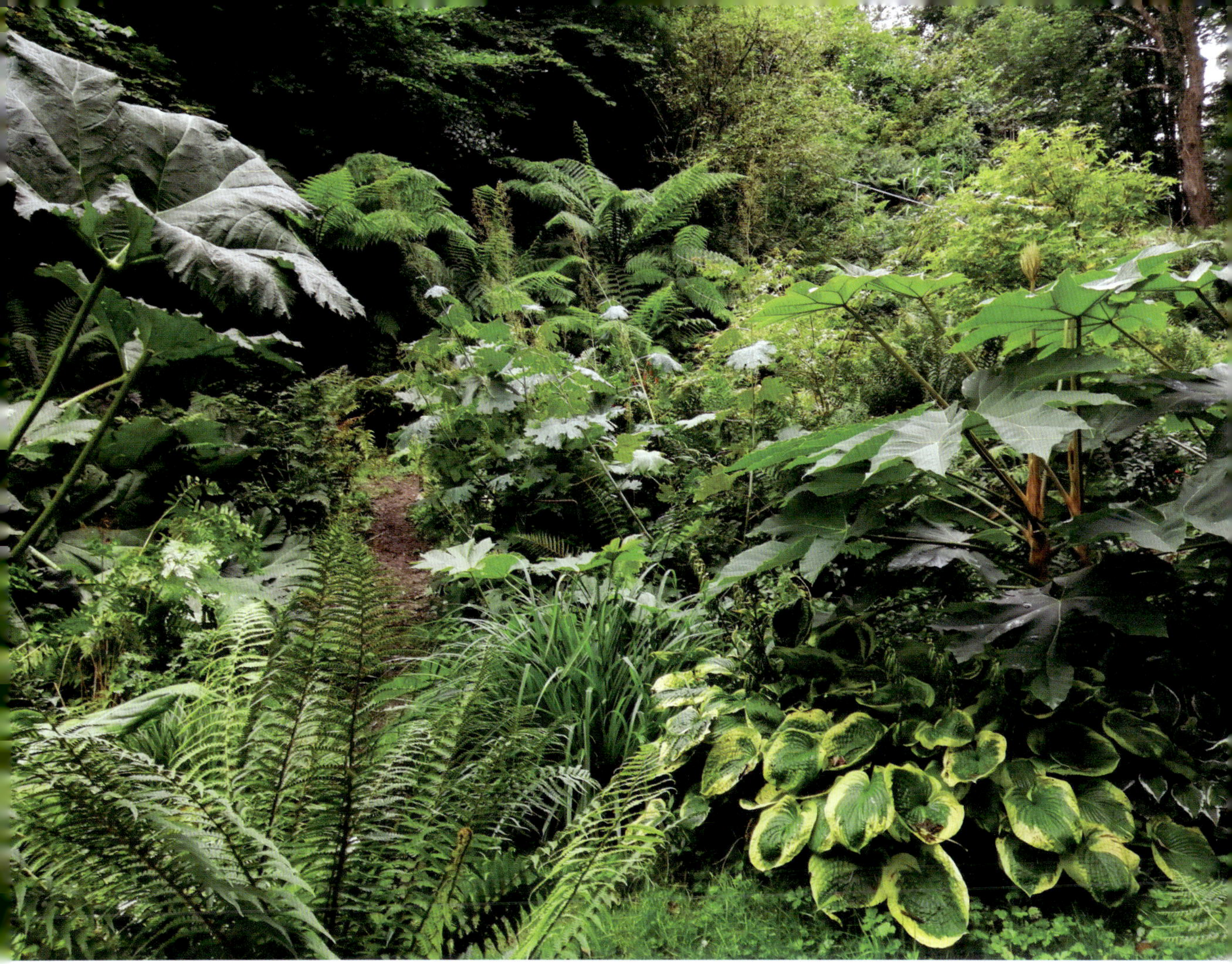

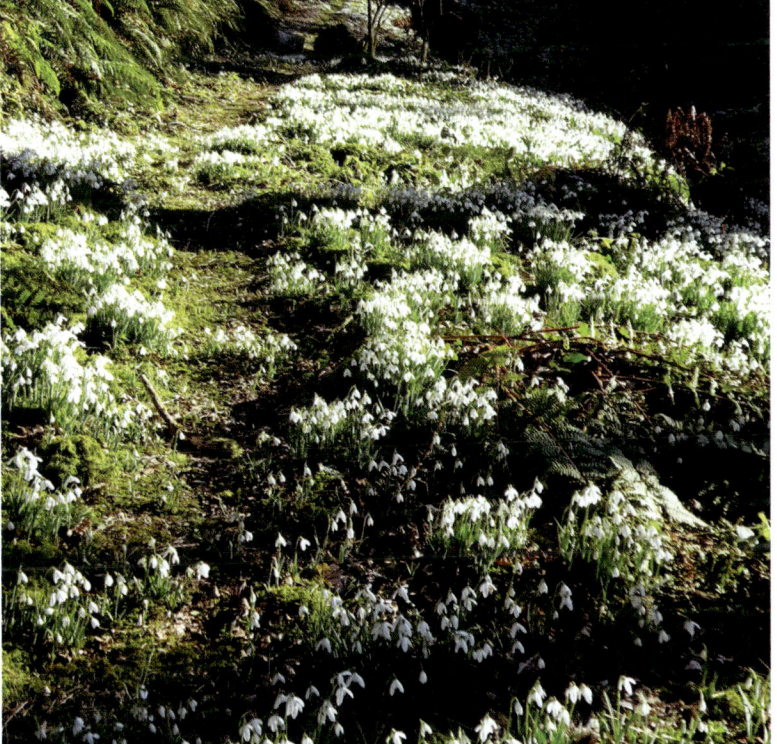

Above.
Large-leaved plants in the ravine.

Left.
In early spring the ravine is a river of snowdrops.

garden. John is, of course, adept at pulling out the nettles which still appear here and there in the walled garden, and I am happy to help gather up armfuls of sticky willie (cleavers) that swamp the ferns in the ravine in summer, and to cut away the dead fern fronds in winter to lay on the paths, keeping them free of mud.

The sunken garden

The sunken garden is the most recent addition, bought as part of the package that included the north half of the walled garden. It sits almost 4 feet below the road, which may have been artificially built up when the farm buildings were demolished in the late 1950s, in order to make a track to the newly constructed Barholm House beyond the Castle. It was also covered in brambles and nettles and had a fair amount of buried plastic rubbish in it when we started work on it. There were two nice rowans, an apple tree and a laburnum, which we were able to save, once the ivy had been stripped away and the dead wood removed. Last year I had chipped bark paths constructed, bordered by fallen stones, a walled niche for a curved stone bench and a couple of small rockeries beside two sets of steps that were made leading Yesdown from the road.

This is a shady garden, overlooked by the rowans, apple tree and laburnum on the road side and an ash tree growing against the wall of the walled garden. The ash shows some evidence of dieback, but it is not clear how long

The sunken garden in spring, with the old piggery in the background.

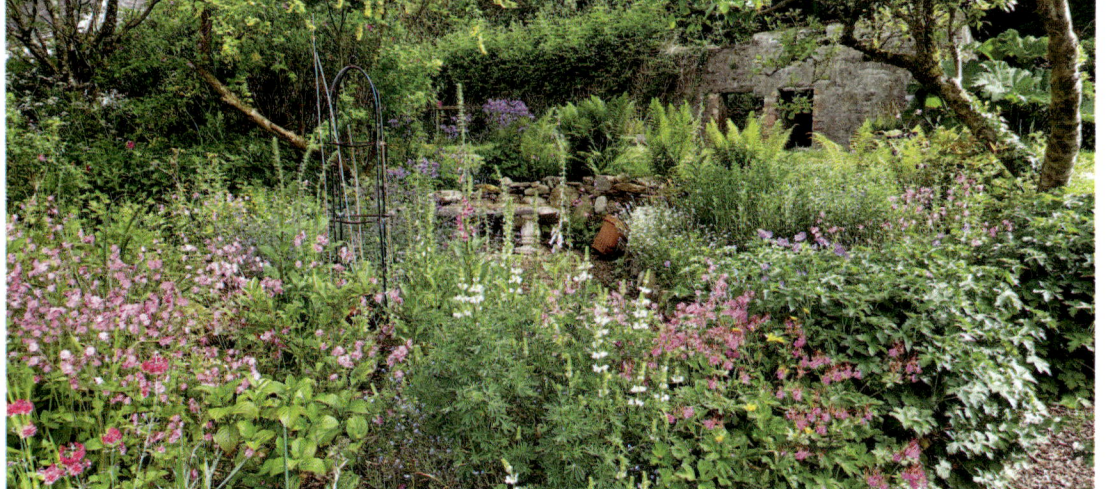

THE DEVELOPMENT OF BARHOLM CASTLE GARDEN

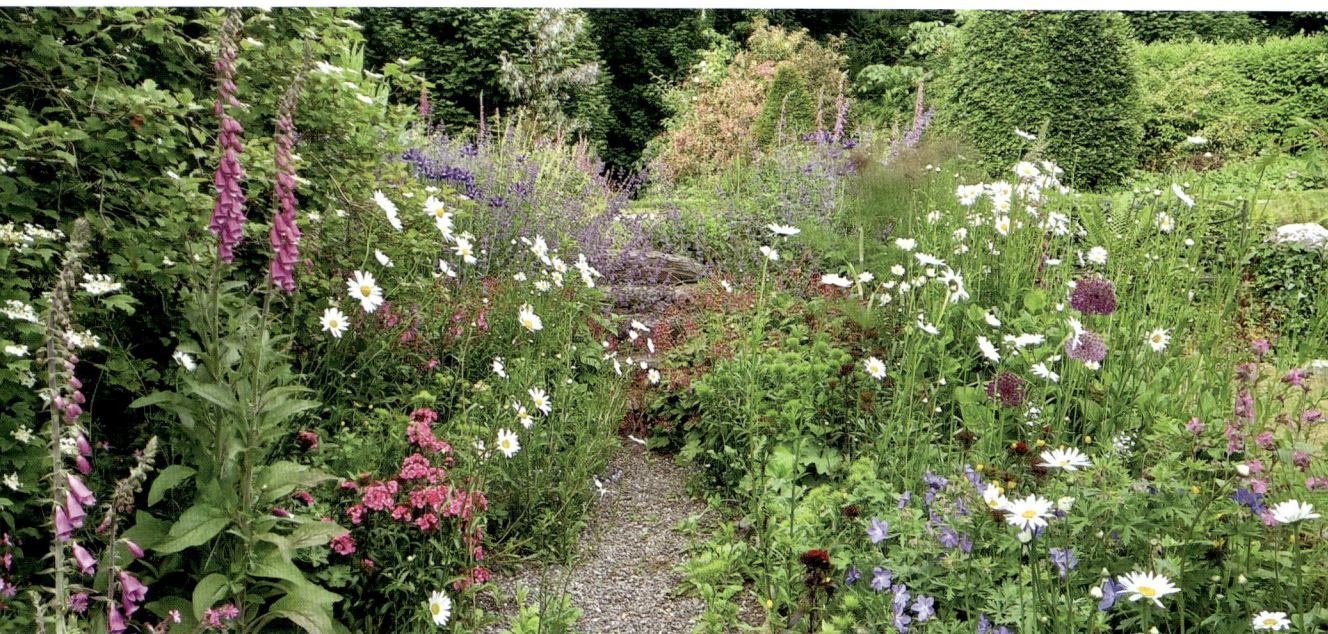

Steps down into the sunken garden from the road.

the disease will take to kill the tree, so for the moment we carry on as if its life is eternal. In spring there is, of course, very little shade from these deciduous trees, so the daffodils (already there, probably from forty years ago), rockery plants and primulas enjoy sunshine when it comes. In the two main beds, mulched with chipped bark, I have planted a tree fern, hostas, pulmonaria, hardy geraniums, hydrangeas, candelabra primulas and colourful mixed heucheras, all of which like dappled shade and are happy there in the summer months. Foxgloves, ox-eye daisies and hardy geraniums thrive in the sunnier parts.

The castle rock

The castle is built on bedrock and surrounded on the south and east sides by rocky outcrops, on which are planted various shrubs and rockery perennials. This area is accessible to all passing rabbits, so it has been very difficult to establish plants, and many have been nibbled to death. My favourites are the weeping silver birch 'youngii' and several hydrangeas. The purple *Cotinus*, which is flanked by two *Spiraea arguta* 'bridal wreath', makes a good splash. A *Chaenomeles* 'crimson and gold', set against the dyke, flowers from Christmas until May then produces (inedible) golden quinces in late autumn.

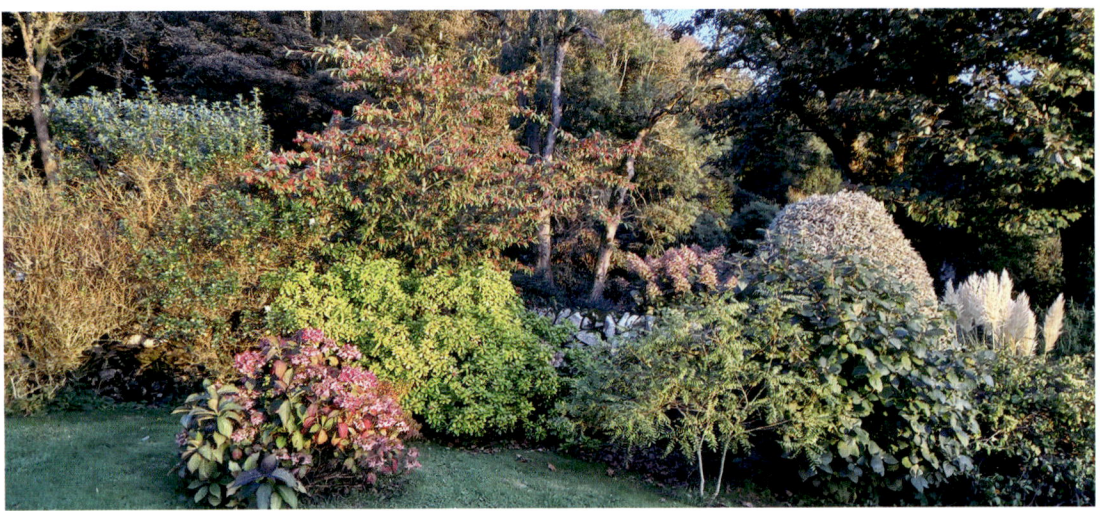

Autumn on the castle rock, where the soil is thin over the bedrock.

Managing the Garden

Garden thugs

The garden thug is an interesting concept. The Royal Horticultural Society (RHS) offers an extensive list of garden plants that have the potential to run amok – but it depends on the situation and conditions. Thugs that I have planted and would welcome more of include *Acanthus*, *Physalis*, *Houttuynia*, orange *Alstromeria* and Russian vine (even although the Russian vine is called the 'mile-a-minute plant' its performance in our garden has been so sluggish as to amount to no more than a few inches each year; the ferocious winds in the paddock keep it stunted). Thugs that I have planted and now have to battle with their spread include Japanese anemones (which march across the lawn and have to be mown down), white rosebay willow herb, *Chamaenerion angustifolium* 'alba' (don't believe claims that it is less vigorous than the common pink variety), *Alchemilla mollis*, *Sorbaria*, *Macleaya*, yellow loosestrife, *Doronicum*, *Vinca minor* and *Rubus*.

The *Tetrapanax papyrifer* 'Rex' (Chinese rice-paper plant) is a magnificent, exotic-looking (and expensive!) plant, but we weed these out, as they sucker all over the ravine, and have now started to spread their way across the paddock and the walled garden. I have two enormous ones in the sheltered half of the walled garden, which make statement specimen plants at 10 feet high. In the paddock, they have adapted to the windswept conditions and remain at knee-height. The stag's horn tree (sumach or *Rhus*), which originally came from nearby Cally Gardens, took a long time to establish but has now produced hundreds of tiny suckers, which we just mow over to get rid of, leaving a choice few for structure.

Arum italicum was already widespread in the new part of the walled garden and, although it can be a bit of a nuisance, it does compensate with marbled leaves and a colourful display of bright orange berries in late summer. Unfortunately for us, the birds have discovered the berries, so these rarely last more than a week or so. The only plant I am fearful of planting is the bamboo, so I have none in the garden; I have no wish to add to the list of thugs needing control.

The plant thugs also serve a purpose in adding to the stock we sell to garden visitors and pass on to the local community centre and plant sales. All of the proceeds go to charity and fulfil the dual purpose of making people happy and raising money for good causes. What could be better?

Weeds

A weed is just a plant in the wrong place. We are surrounded by thousands of acres of uninhabited countryside where the native nettles, docks, etc. are able to thrive and provide food for insects and other beasts. In Barholm garden, we do not welcome coarse grasses, nettles, docks, dandelions, buttercups, bindweed and ground elder into the cultivated areas. Since I use no weedkillers, they have to be pulled, dug or hoed out by hand. The best advice, which it is not always possible to follow, is to hoe lightly whenever the weather is dry. That cuts the weeds off from their roots and kills annual weeds such as chickweed, hairy bittercress, fine grasses, etc. Some weeds do not respond to the hoeing cure. Lesser celandine is so adaptive that hoeing simply helps to spread its little bulbils and creates a worse problem. It is pretty in early spring, with its cheerful yellow flowers and bright heart-shaped leaves and is a good source of early pollen and nectar for pollinating insects. It soon disappears from sight once flowering is over, so I leave it. Hoeing dandelions and dock is only a temporary measure, as their cut-off roots will send up new shoots pretty quickly. But persistence pays and constant hoeing off will eventually kill the rootstock. I have lost the battle with ground elder in certain beds, and just let it be. The summer flowers are attractive, rather like Queen Anne's lace. Ladybirds love them. Apparently, the leaves can be eaten but are not particularly tasty.

I see weeding as an integral part of gardening, not as a great nuisance. It gives an opportunity to get close to the soil and keep an eye on what is going on at ground level. I am no longer able to kneel, but I can still bend at the waist and get on with weeding that way. If you weed every day, it is easier to keep on top of things, of course. And if you have the time, resources and energy to mulch the soil deeply, any time from autumn to spring, that will help to suppress the weeds quite effectively and also give an opportunity for the worms to pull down material into the depths and make the soil below more fertile.

All weeds are wildflowers, and some are especially welcome at Barholm. The ravine is carpeted in dog's mercury, which is an indicator of ancient woodland, the designation of Barholm woods. Apart from the fact that it is highly poisonous, it is an innocuous plant with insignificant little flowers on the male plants which provide a green underlay to support the ferns and also the beetles that feed on it. We welcome wild garlic, or ransoms, in the late spring, enchanter's nightshade, hawkweed, fumitory and bluebells. Ground-cover plants are a really useful means of weed suppression. I use spotted deadnettles, *Lamium maculatum*, *Doronicum*, *Persicaria* and, of course, ivy to blanket the ground in new beds.

Biodiversity

What we offer to the birds, bees, animals and insects is biodiversity; there are many hundreds, probably even thousands, of different varieties of plants in Barholm garden. Many are not native, but nonetheless they are a vital source of pollen and nectar throughout the year for the bees and insects that keep our plant life alive. Indeed, more than 90% of plants in British gardens are non-native; we would be lost without introduced species. As I write this, in February, the garden is full of flowering non-native snowdrops, with a supporting cast of crocuses, aconites, daffodils, camellias, quince, mahonia and periwinkles, all non-native. Almost every spring-flowering plant in our gardens has originated from outside the UK.

Red squirrels visit our bird feeders every day, along with great spotted woodpeckers (who peck at the lime harl surrounding the window), nuthatches, coal tits, great tits, blue tits, chaffinches and the occasional field mouse. Overhead, the predators circle – buzzards, ravens, crows, sparrowhawks, peregrine falcons and red kites. At ground level, the fox constantly harried our neighbours' chickens, until they finally gave up keeping poultry. Mr Mole travels widely across our garden, leaving tracks and hillocks as

THE DEVELOPMENT OF BARHOLM CASTLE GARDEN

137

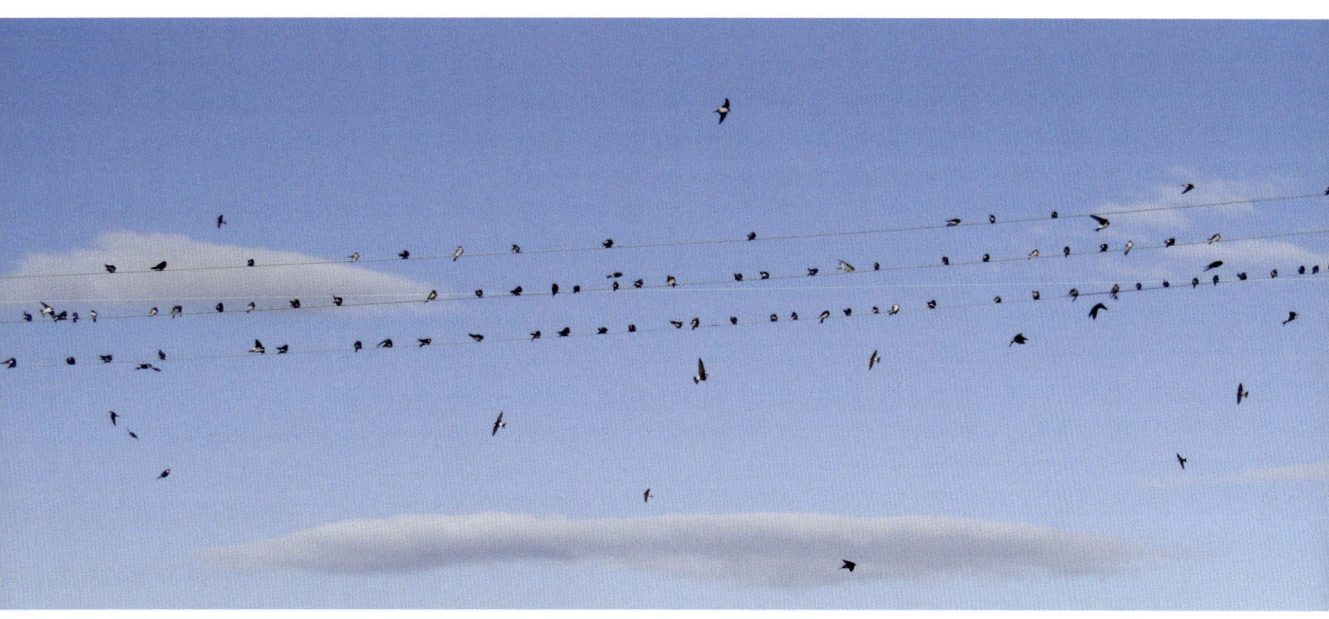

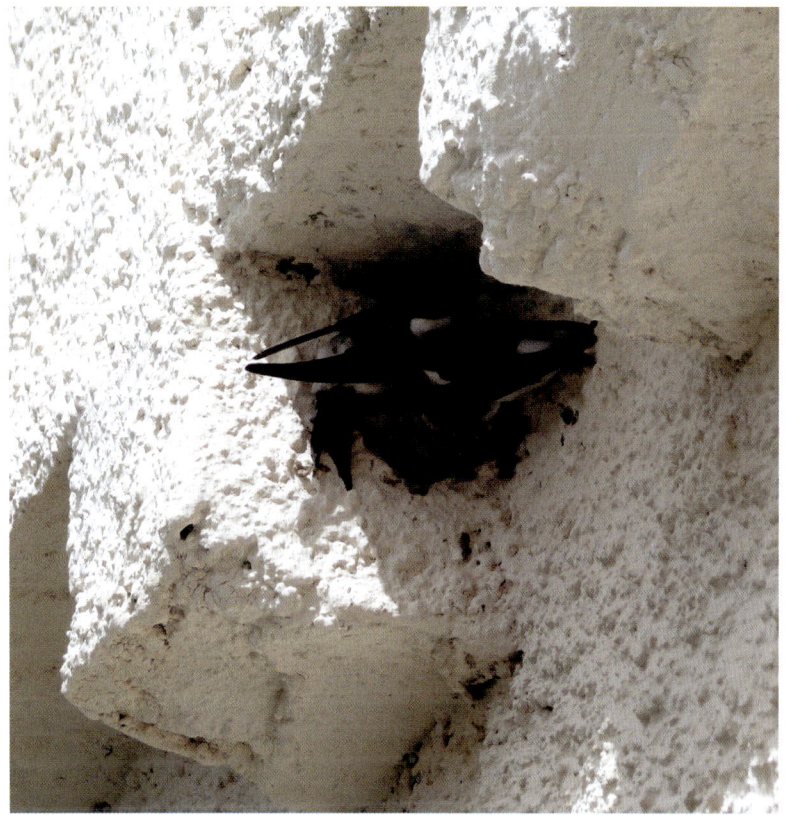

Above.
House martins congregating on the wires above the walled garden.

Left.
House martins building their nest in the crenellations of the stair tower.

evidence. I use the earth thrown up into the molehills as handy potting compost. Together, the cast of living creatures makes a great garden community, even if some are more welcome than others.

Before starting to plant the paddock garden we had the central telegraph pole taken away and the overhead cable sunk (at a large cost) so that we could plant trees without worrying about height. We had not thought of doing this in the walled garden, which was the first part we cultivated; however, there are no tall trees in the centre and the joyful thing is that the telegraph lines provide a place for our visiting swallows and house martins to perch and congregate. It is a real pleasure to watch the birds on the wires; we are so glad they are there. The house martins arrive mid-April and make their nests in the corbelling of the stair tower. We can view their comings and goings from the little window in the corridor outside our bedroom, until they leave in late September.

Rabbits and deer

Not every living creature is equally welcome at Barholm. From the start, the rabbits were a major feature in the garden and I was dismayed at the number of lovely plants which I introduced, only to find them completely eaten within

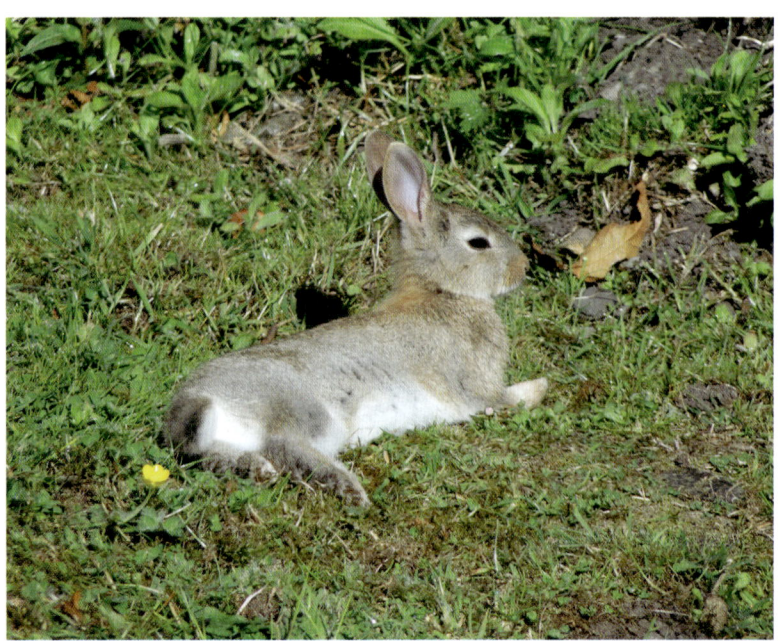

Indolent rabbit, lolling about disdainfully.

days, or even hours. The population of rabbits seemed to explode in the first few years, so we decided to rabbit-proof the walled garden and the paddock. It was an expensive fool's errand. The walls are drystone dykes, and little rabbits can easily wiggle their way through tiny gaps between the stones. Bigger rabbits can actually jump over quite high dykes if they are in a hurry, as I have observed. In desperation, I collected hair clippings from my hairdresser, shaved strips of old bars of soap, bought a systemic solution called Grazers and even resorted to vile-smelling urinal blocks to spread on the soil in order to try to keep them at bay. In the end, only two things worked: one was the terrible disease myxomatosis, which wiped out most of the rabbit population in certain seasons, and the other was simply time and the buildup of mature planting. Rabbits are programmed to investigate recent planting, presumably in case something novel and tasty has been provided.

We have relatively little trouble from rabbits now, although they are still in the garden in some numbers and they still have a taste for stripping the bark from young trees and shrubs in the paddock garden, where I have to protect any new plants with wire cages and bark wrappers. I know that if I were to start a new bed in the paddock and plant it out with tasty little perennial plants, they would very quickly be eaten. In bad 'rabbit years' I even have to place my summer display pots on tables or stands to protect them.

The deer have been less of a problem, although they ate 4 feet off the top of an expensive magnolia in the ravine and they regularly prune back the roses along the side of the track. We keep the paddock and walled garden gates shut from dusk every day, and although a deer could probably jump over if it wished, they are creatures of habit and just keep up and down the track without deviating. The magnolia recovered very quickly and flowered prolifically the next year, as plants are wont to do if they receive a major threat.

Visitors from the city find the rabbits fascinating. I was once in the garden talking to an architect from Glasgow who had come for a Doors Open event, when a rabbit nonchalantly strolled across the lawn in front of us. She stared in astonishment. 'Is that YOUR rabbit?' she asked. I explained that while it lived in my garden it wasn't my personal rabbit, but a wild one. 'But do you feed it?' she persisted, and I told her that it fed itself very well, by eating my choicest plants, especially the tasty little new shoots. Just then, four of next door's hens made an appearance, having escaped from their enclosure, running across our grass, and the architect whipped out her camera in delight. She and her family later wrote to say that, while they had enjoyed the tour of the castle very much, the rabbit and chickens had really made their day.

Other 'pests' and diseases

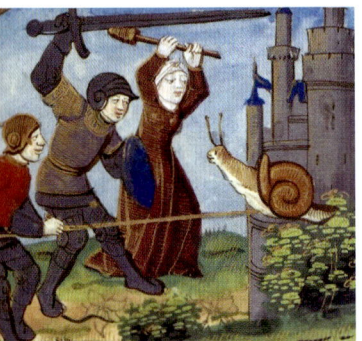
Mediaeval castle pest control.

Like every other gardener we suffer from the predations of slugs, snails, earwigs, etc. The RHS has recently banned the use of the word 'pest', recognising that insects, spiders, slugs and snails play an important role in protecting biodiversity. I have never used any kind of insecticide in the garden and just accept that these creatures have an equal right to live here. Sometimes it is a challenge, especially when your hostas look like shredded lettuce, but the following year usually brings a change. We have bad mole years, mice years and rabbit years, but they are usually followed by years when the populations are lower. At least we have had no badgers so far – they can be severely destructive in gardens. The ideal is that every creature in the food chain keeps the environment in balance – the sparrowhawks eat the small birds, the small birds take the ladybirds, the ladybirds take the aphids and so on. It would be very convenient if it all worked smoothly like that, giving us a garden free from slugs and aphids, but unfortunately there are always thousands of these left at the bottom of the food chain, ready to eat the garden plants.

Diseases do not cause us too much of a problem at present, although the reason we have very few elms is because of Dutch elm disease, which killed the majority of mature trees some years ago. *Phytophthora ramorum* (a highly destructive water mould) has meant the felling of thousands of larches in this area; fortunately, we have mostly escaped its effects so far, although we have lost a few mature *Viburnum bodnantense* and a treasured *Hamamelis mollis*. Now we have ash dieback, which is slowly killing our younger ash trees. Rose black spot is a disease one can live with, unsightly as it is, and powdery mildew is a temporary and counter-intuitive hazard of dry summers.

Compost

In 2017 I finally decided it was time to tackle the most horribly scruffy area of the paddock. In a far corner, the builders had dumped several tons of stone in a large heap that had subsequently been colonised by brambles and nettles; beside it, John P. had used the area for years for dumping the grass cuttings from the paddock. It had become an enormous squelchy mass. We hired a digger to rearrange the huge stones to make a rockery and Jim took away the worst of the squelch, then we filled in between the stones with the drier stuff. It made amazing compost. I stood back and watched as plants romped away as if on steroids. I had to remove a triffid-like fennel which had

THE DEVELOPMENT OF BARHOLM CASTLE GARDEN

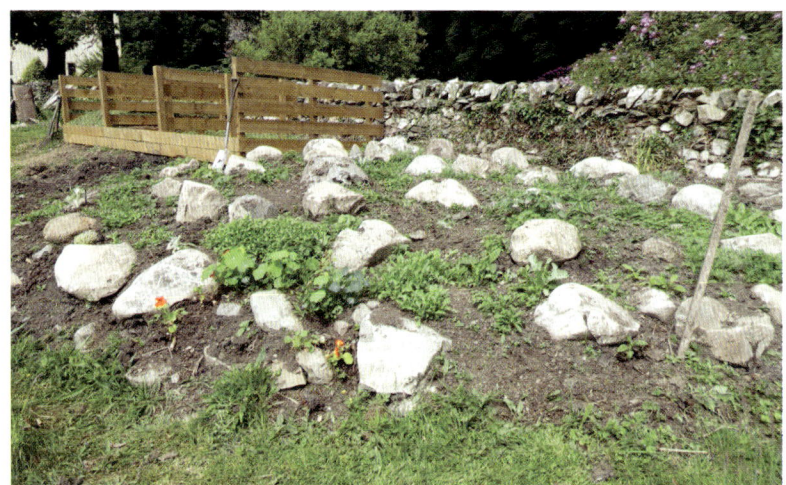

Left.
The rockery under construction, with the three wooden compost bins in the background.

Below.
The first rockery we built.

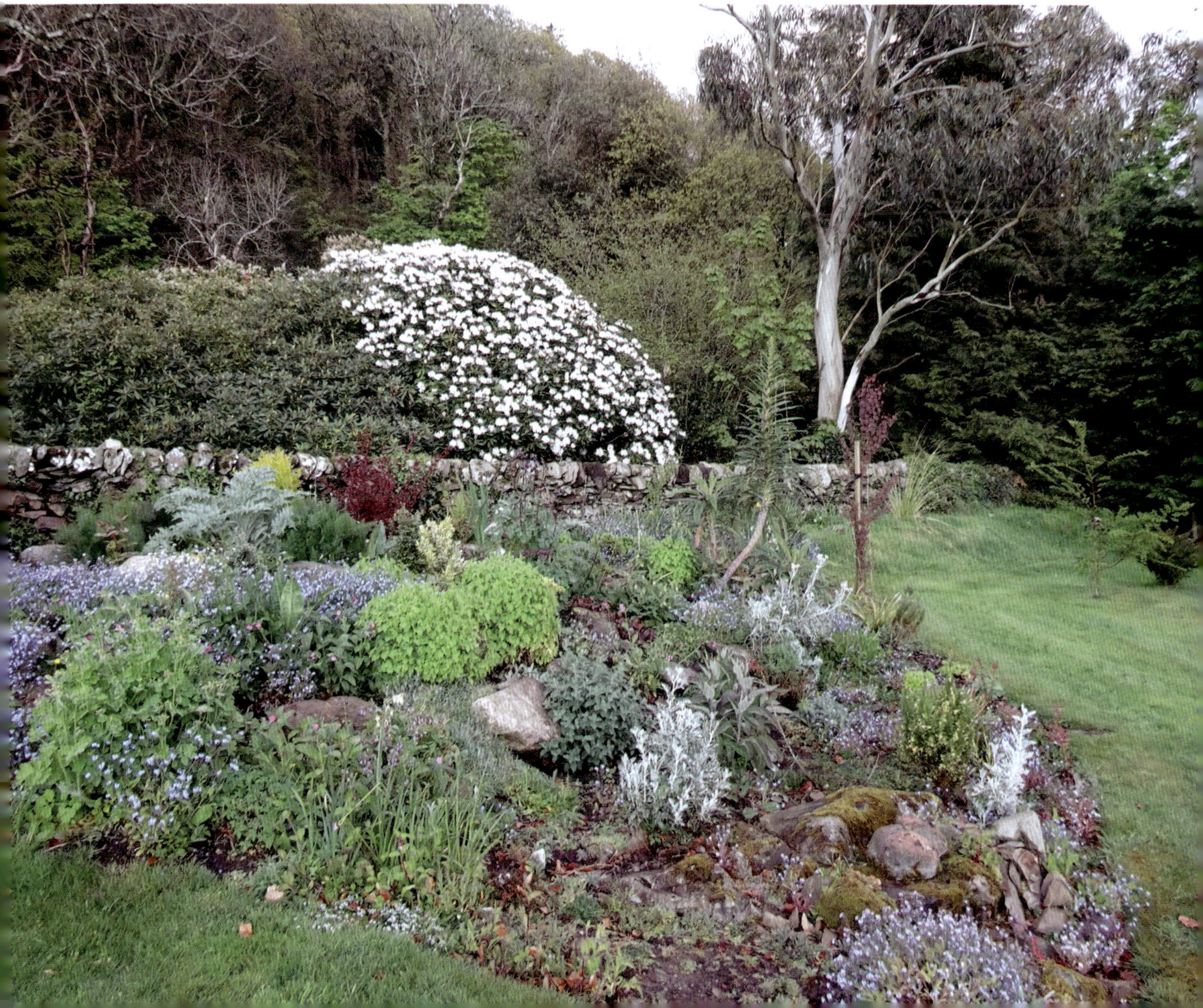

grown much taller than me with dozens of stems thicker than my thumb. I also foolishly thought a climbing rose would do well and look good along the back wall. Once it had rambled all the way through the rockery at speed, I removed it to another part of the garden where it has behaved with much greater moderation.

Beside this area of rockery John P. built three large wooden compost bins, the size of double beds, with wooden slats – a gardening dream come true – and I instituted a proper system of layering grass cuttings with fibrous garden waste, filling one bin each season before covering it and moving on to the next. The compost is of fantastic quality, especially if it is left for three years to mature. We don't have the resources to turn the huge heaps, so they are left to rot down over time.

We all know that we must no longer use peat-based products for potting, but most gardeners, like me, have struggled to find a satisfactory commercially produced alternative. I have started to make my own potting compost, sieving three-year-old compost from the heap and mixing it with sharp sand and very fine grit. My compost is not sterilised, as it should ideally be, which means that weed seedlings do spring up, but when they germinate in pots, small weeds are easily pulled out. Although I also try to keep pernicious perennial weeds – docks, dandelions, ground elder, brambles – out of the compost heap, inevitably some are missed and they will just have to be pulled up again.

Water in the garden

We are not on mains water at Barholm. Our water comes from the hills, via a large holding tank next to the road behind us and an ultra-violet treatment unit in the boiler house. Our water in the house is thus safe to drink, but the garden water is not sterilised. We have six strategically placed external taps with hoses attached which cover the whole garden when necessary, and I collect rainwater throughout the year in faux-lead containers placed under the mediaeval water spouties, which we have in place of gutters and drainpipes. These containers are very useful for dipping and soaking pots of shrubs and flowers. In addition, John rigged up a system to collect rainwater from the shed gutters in barrels, which can be drained to run via an underground pipe down to the large pond in the paddock garden. So, along with the water that regularly comes down in large quantities from the heavens above, we are well supplied to keep the garden going.

It is not possible to set up a sprinkler system with our low-pressure water, so we never water the grass, which has to take its chances. It is very rare that we have a long enough period of drought to affect it, and if it does turn a little brown occasionally, it soon recovers when the rain comes back. Most of the plants in our garden manage perfectly well without additional watering *except* when they are newly planted or are in pots. Because I am constantly planting and moving trees and shrubs as well as annuals and perennials, and I have around 100 ornamental pots and planters in the garden, I do need to have a ready supply of water. The stream in the hills that supplies us and our neighbours has never been known to run dry; luckily, we do not have the problems that many people on mains water in England suffer.

Grass and lawns

When we first opened our garden for visitors as part of the Scotland's Gardens charity (the 'yellow book') scheme in 2012, participating gardeners were expected to offer interesting planting and – most importantly – a tidy and well-maintained garden. We did have the interesting planting, but I felt rather ashamed of our overgrown corners and steered visitors well away from them. Now, happily, attitudes have changed: gardeners are encouraged to provide plenty of wilder areas for pollinators. I almost feel we have to apologise for cutting the grass at all, but our lawns are far from sterile green deserts. The blackbirds, pigeons and thrushes in the walled garden benefit from feeding in shorter lawns and there are still many small wildflowers amongst the grass for insects – clover, speedwell, selfheal, daisies and cinquefoil. We do still have lots of overgrown corners, but now they can be recognised as areas for the protection of garden wildlife, where insects and small mammals thrive.

Our area of grass is inevitably large in a garden of 3 acres. In the walled garden we have three lawns and they are mown once a week, in keeping with the semi-formal layout there. In the paddock garden the tractor mower is used on the main area of grass every two weeks, and a prodigious quantity of clippings comes off each time. It is all added to the current compost heap, where it is lasagne-layered with coarser garden waste to give us beautiful crumbly compost after a couple of years.

We did try using a robot grass mower for a while, but although it was very entertaining to watch, and did a good job when it was actually working, there were a few major drawbacks which meant it was not for us. The set-up involves burying guide wires along the boundaries of where the mower is to roam.

This was a very complex business, as the walled garden has island beds and wall beds and differences in levels. However, Jim managed it for us and all was well until the mice, rats or rabbits discovered the underground wires and nibbled through them in various places. Finding the fault was a nightmare and as fast as we made joins in the broken wires, more breaks were chewed in unspecified places. It also became obvious that my garden is simply not static. I am constantly expanding beds and making new ones; every time a change happened, the wires had to be re-routed. A further problem was the garden's changes of level; the robot did not cope well with the slope towards the path, sometimes making it up to the top, only to teeter precariously and fall head over heels back down. I would recommend a robot mower only if you have a regular lawn with no plans to change its shape and no gnawing animals living below the surface.

The greenhouse

We had our wonderful greenhouse built in 2011, just before we moved in to Barholm Castle. It is capacious and, thanks to the demand of the planners that we must install cavity wall insulation (in a building that is almost entirely glass and rarely heated), there is a broad windowsill all around the edges, which is perfect for holding plant pots. In the greenhouse I grow mainly

The greenhouse, built in 2011: interior.

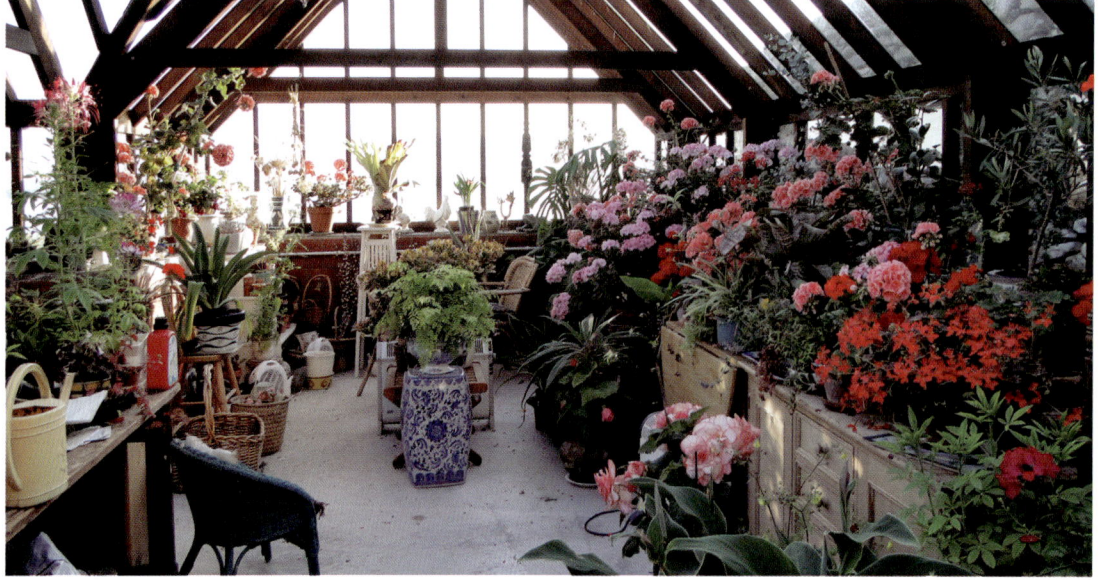

THE DEVELOPMENT OF BARHOLM CASTLE GARDEN

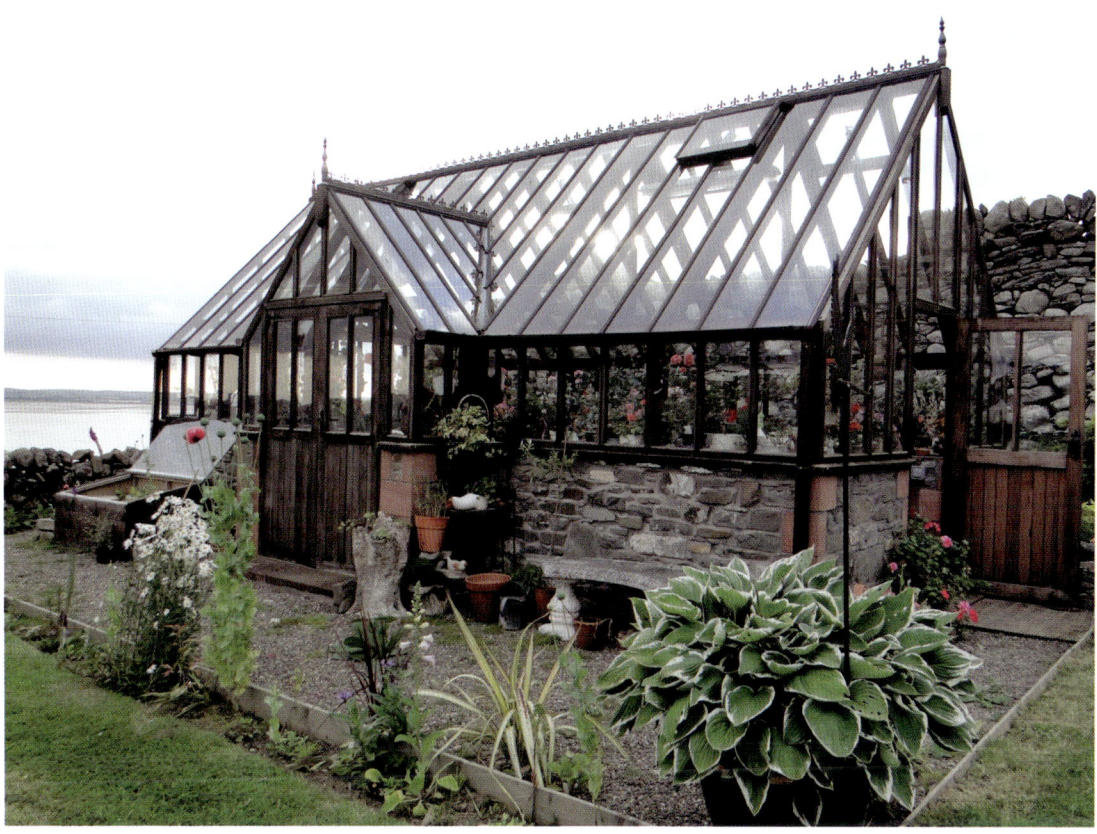

The greenhouse: exterior.

succulents, cacti and various houseplants, including lots of geraniums, treating it more as a conservatory, although I also propagate lots of seeds in the spring. Inside the castle there are not enough windowsills with direct light to keep my large collection of houseplants satisfied. Most of the greenhouse plants need watering daily in the summer, so that is my first task of the day every day from May to September.

Propagating

When the garden was developing at first, every plant was precious and a potential source of new stock. I still do a lot of propagating, mostly by division, placing the new plants elsewhere in the garden. Chopping a piece off the edge of a clump of perennials, or even pulling a bit up (by accident or design), is by far the easiest means of increasing your stock. Every spring, I grow a number of plants from seed in the greenhouse, mainly easy half-hardy annuals such as cosmos, lavatera and amaranthus, and sometimes perennials, such as tree lupins, from saved seed. I also propagate some seeds on the kitchen floor, using the underfloor heating to provide enough gentle warmth for very effective germination. LED propagation lights have become much cheaper recently and are easy to install in propagating frames, overcoming the lack of natural light in the kitchen. Greenhouse succulents are mostly simple to propagate, usually producing new roots in water after a couple of weeks. Some kalanchoes (*diagremontiana* and *pinnata*) produce dozens of plantlets along their leaf edges and these drop off into neighbouring pots, producing new plants. They are evocatively named as Mexican hat plants, mother of thousands, cathedral bells and devil's backbone, and they sell well at charity plant fairs.

When I made the first island bed in the paddock, I decided to sow a summer wildflower patch in the newly dug soil. Accordingly, I bought a large pack of an expensive annual native seed mix and carefully followed the protocol for sowing: raking and destoning the soil to a fine tilth, measuring out the patch into square metres marked by canes, weighing the seed into equal portions and mixing it with coloured sand, then broadcasting seed into each section, tamping down and finally watering with a fine spray. It took a lot of time and effort but looked very satisfactory and I anticipated a beautifully bejewelled patch in eight weeks' time. However, that night my plans were

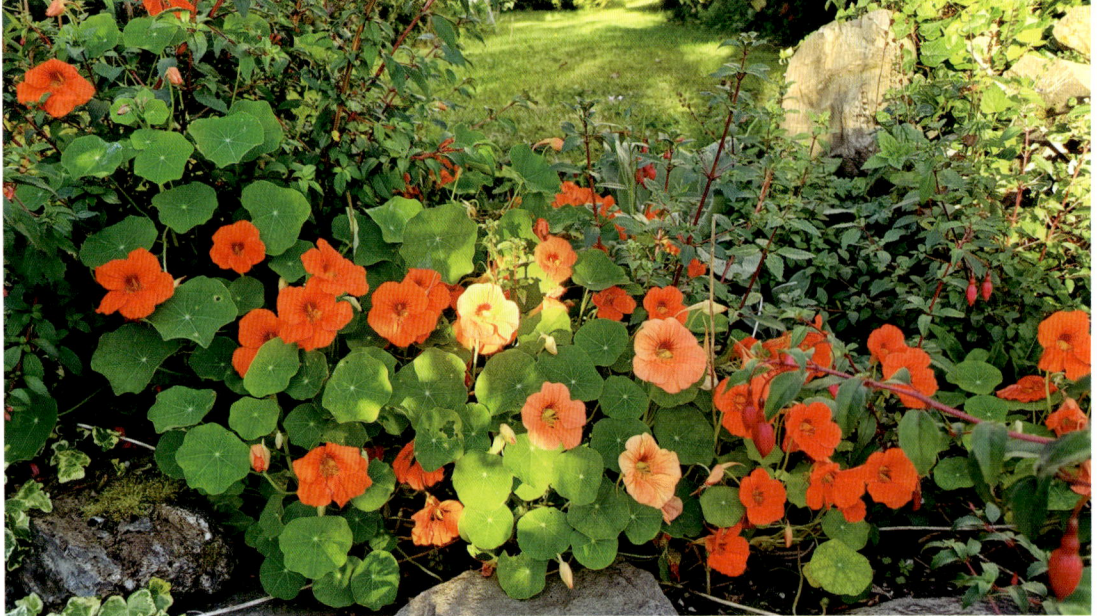

Nasturtiums in the walled garden, grown from seed.

thwarted, as a storm of biblical proportions swept across south-west Scotland, the thunderous rain washing away all of the loose topsoil in rivulets and my expensive seed with it. Since then, I have adopted a much more cavalier approach to seed sowing. I use the packets of seed that come free with my weekly gardening magazine, and if I see a patch of bare soil anywhere in the spring, I randomly scatter some seeds there while it is raining. As a technique it is a bit hit and miss, but I do get some great flower displays out of it. And the bonus is that these plants often go on to set seed which germinates afresh next year, with no effort on my part. And so the cycle of gardening goes on.

Fruit and vegetables

There seems to be a moral imperative to the growing of garden produce. It is simply expected of anyone with a large garden. In 2020, I eventually caved in under pressure of questions about the location of the non-existent vegetable patch. I use the two fallow compost heaps to grow several types of vegetables. They act like super-charged raised beds. I must admit that my favourite crop is the pumpkin – the crazy speed of growth makes me laugh; they even climb up nearby trees. I use them as ornaments piled outside the front door in autumn, to the disapproval of those who feel that pumpkins should be made into soups and pies. Other crops are beset with problems. Potatoes get blight, lettuces bolt, cabbages and beans attract nibbling mice, tomatoes split and courgettes over-produce. Perhaps I am just not committed enough to success, although I have great admiration for those who produce abundant edible crops. Watering is a nightmare in dry weather, to add to the stress. However, we do dine well on home-grown potatoes and courgettes most summers, as these are almost fail-safe crops.

Herbs and spices are much easier. I grow various varieties of mint – never sufficient for cooking in with the potatoes and courgettes, despite letting it run unrestricted – bay, fennel, hyssop, lovage, oregano, parsley, rosemary, sage, Sichuan pepper and thyme. Most of these I use in day-to-day cooking. I also have a rampant hop plant which comes back every year and makes interesting dried garlands in autumn.

Fruit is much less problematic than vegetables. The old apple trees in the walled garden, probably planted in the 1970s, are bountiful, and the new ones I have planted are good at producing apples now. The two old bullace (wild plum) trees in the walled garden give a very pretty crop, and the fruits are sweet enough to eat straight from the tree. We have figs, rhubarb, rasp-

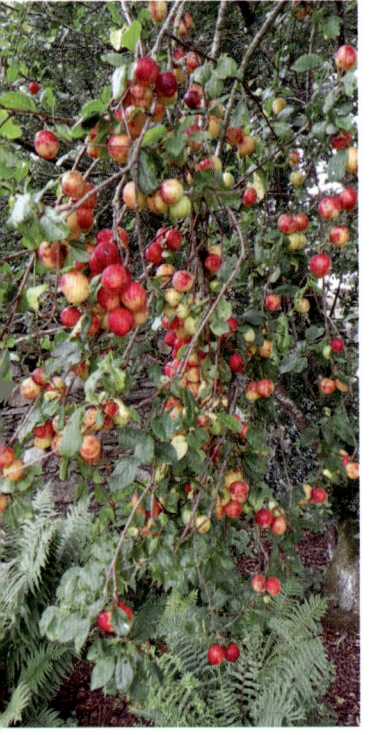

Wild plums in the walled garden.

berries, blackcurrants, gooseberries and wild strawberries but I don't harvest them all, just gratefully gather a few that the birds have left for us. My favourite fruits are the plump brambles that line the back road behind the castle; every autumn I make apple and bramble crumble. The apples that we don't pick or give away provide a feast for the birds and mammals in the garden, and windfalls usually last until the weather gets colder in November.

Plants for free

We have so many plants for free in our garden: self-seeders are a gift to the gardener. These plants fill gaps and are wonderful accessories in our overall aim of keeping the show going, filling the beds with colour and suppressing the weeds. The Americans have a good word for self-seeders: 'volunteers'. Our volunteers are the supporting cast of the garden at Barholm; they include honesty, hypericum, leucanthemum, feverfew, hawkweed, campion, forget-me-nots, foxglove, opium poppies, primroses, caper spurge, self-heal and Welsh poppies. This list is of the ones that I did not plant in the first place – they just came along with the birds or the wind. Others that I *did* plant, which now spread happily, include aquilegia, *Erigeron karvinskianus* (Mexican fleabane), sweet williams, pulmonaria, hardy geraniums, fuchsias and furry-leaved lambs' lugs (*Stachys byzantina*). A few give me only the occasional seedling – a francoa has appeared growing out of a low wall, to my delight. They grow in rock crevices in their native Chile, so must feel at home here. *Phygelius*, or cape fuchsia, drop seedlings near the mother plant, again in awkward crevices. And I am nurturing a couple of physocarpus seedlings that have appeared in a pot. Pampas grasses self-seed in unusual spots, doubtless spread by the sparrows who love to feed on the seedheads. They grow huge very quickly, so they need to be restricted to the wilder parts of the paddock.

Trees and shrubs that self-seed or spread in the garden include beech, berberis, birch, buddleja (both purple and white), cherry, cotoneaster, hawthorn, hypericum, leycesteria, oak, rowan, sumach and *Tetrapanax*. Mostly I let them grow where they appear, or if inconvenient I move them to a better place or pot them on as gifts or for plant sales. When we replaced an awkward triangular part of the lawn with pea gravel, a cornucopia of seedlings appeared. Gravel is an amazing natural material in the garden. It supresses the really unwelcome weeds (coarse grasses, nettles, docks, buttercups and hogweed) and encourages the plants I want to see. It also keeps the rain off those that are susceptible to winter damp and protects the soil below from drying out in periods of drought.

Flowers

The garden's principal appeal for me is aesthetic. I love to look at, and walk among, colourful plants and flowers. First thing every morning, I get out in the garden with my Jakoti snippers, tidying, deadheading and cutting back. When we lived in the Netherlands, a vast array of cut flowers was available very cheaply in supermarkets and flower stalls and I always had vases filled with seasonal flowers through the house. Now that I have a large and floriferous garden, I again have vases filled with seasonal flowers through the house, but I have grown and picked them myself, which feels very rewarding. Every Christmas Day our table centrepiece is a vase of flowers picked from the garden the day before. Even in the hardest winter I can find a few bright blooms and berries to put in a December vase.

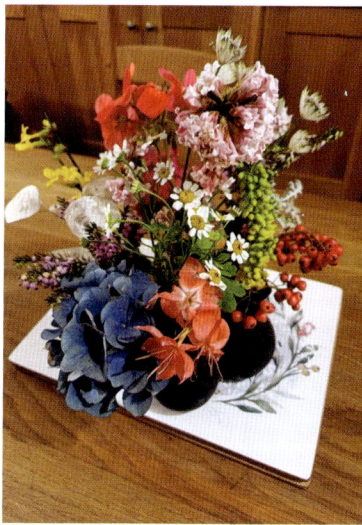

Christmas Day flowers in the kitchen.

Some flowers are very fleeting, but if they are spectacular enough, like peonies and poppies, I am happy to give them space. The very best cut flowers, which I have in vases in the kitchen from May until December, are hydrangeas. They come in an extraordinary range of colours and forms, and they last well in water. An added bonus is that, towards the end of the growing season, a few from each bunch will spontaneously dry out indoors and can be transferred to dried flower arrangements for the winter.

Storm damage

In November 2021, Storm Arwen, an extratropical cyclone, caused significant damage to gardens and woodland across Scotland. We lost a very large sycamore tree which, it turned out, had been hollow inside. It fell across the area beside the road where I had planted six large-leaved rhododendrons from Cally Gardens, four special camellias, an embothrium, a variegated myrtle and some deciduous azaleas, plus various smaller rhododendrons. An old deutzia was also in the path of the falling branches. All of these shrubs were flattened and pinned down beneath tons of wood, but the tree surgeon could not get to us until Christmas Eve, as he was overwhelmed by pleas for help from people who had lost roofs and had had their electricity supply cut

Jim helping to clear the last of the fallen sycamore after Storm Arwen in 2022.

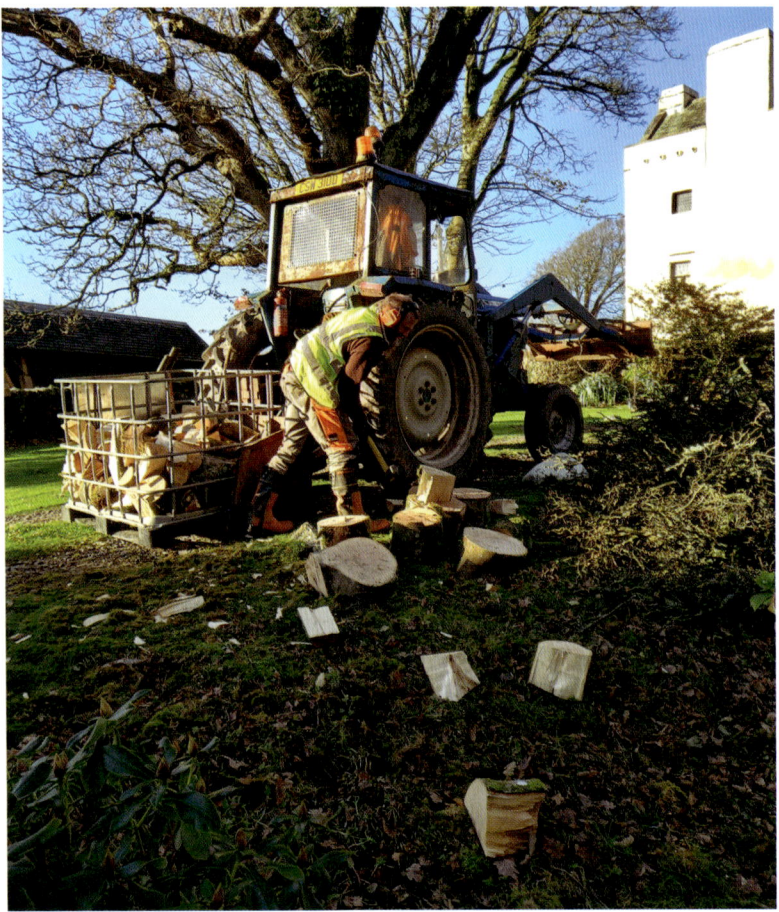

off. It took thirty man-hours to get rid of all of the trunk and branches, and when the area was finally cleared, I looked at my special woodland shrubs and shed a few tears. All were very badly damaged, and I could not imagine how they would ever recover. However, the embothrium flowered amazingly well the next year, despite having lost several of its main branches – when a plant is under threat, it often responds by flowering better than ever, to ensure survival. The big old deutzia had to be cut down to ground level, as all of its branches were splintered and broken, but it also flowered well the next year. The rhododendrons are beginning to recover, although all are growing with a slight lean now, having had the upright bashed out of them, and the camellias look very healthy and have been very floriferous since then.

More recently, Storm Kathleen in April 2024 brought down an oak on the side of the ravine, which then brought down a sycamore and destroyed a large section of our Himalayan path, shredding the handrails like matchsticks and undermining part of the path itself. John constructed an ingenious bridge across the dangerous part and it is now walkable again. In January 2025 Storm Éowyn felled nine large trees and several smaller trees and shrubs, including a precious ozomanthus and my large embothrium. Gardeners are increasingly vulnerable to extreme weather events but, in the end, we recover and live to plant another day.

Garden Structure and Features

Every garden needs structure to support the planting. We have made a number of major additions throughout the garden, including paths, dykes, hedges, ponds and rockeries. We discovered the hard way that any lightweight structure will not be able to withstand the weather, particularly the winds, for long. A large composting bin, which was half-full, blew away in a storm – I found it a few days later, three fields away. A woven willow gazebo lasted less than one season before dramatically collapsing in dozens of pieces. It has been replaced by a solid wooden version, and a second one was added when we expanded the walled garden.

Paths

My favourite path is the crazy paving path in the walled garden. When we ran the landscaping company Cree Landscapes, a client wanted her arts-and-crafts crazy paving deconstructed and pebbles laid in its place. I tried to dissuade her, but she was clear about her wishes. She also wanted the paving stones taken away, and we were only too happy to oblige. Her back garden became our path to the castle door, with large flat slabs laid in crazy paving fashion. They would have been very costly to have specially quarried and they look as if they are meant to be there.

Many of our paths are made of chipped bark from our own trees. Once every couple of years the energy company sends a party of woodsmen to assess whether any of our trees are in danger of interfering with the overhead electricity cables. They usually cut down some major branches, then chip the wood and give it back to us for mulch and paths. When we have a fallen tree logged, we ask the woodcutter to chip some of the branches and use it to make or top up paths.

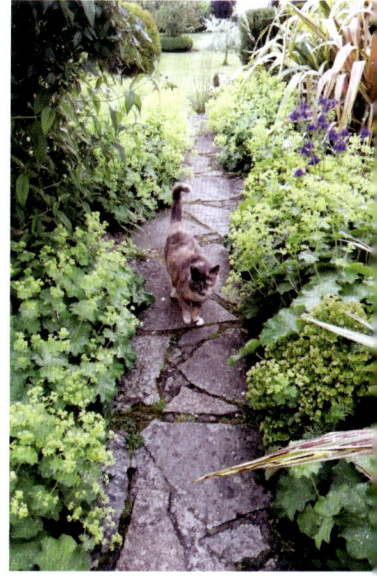

Musky the farm cat strolling up the crazy paving path

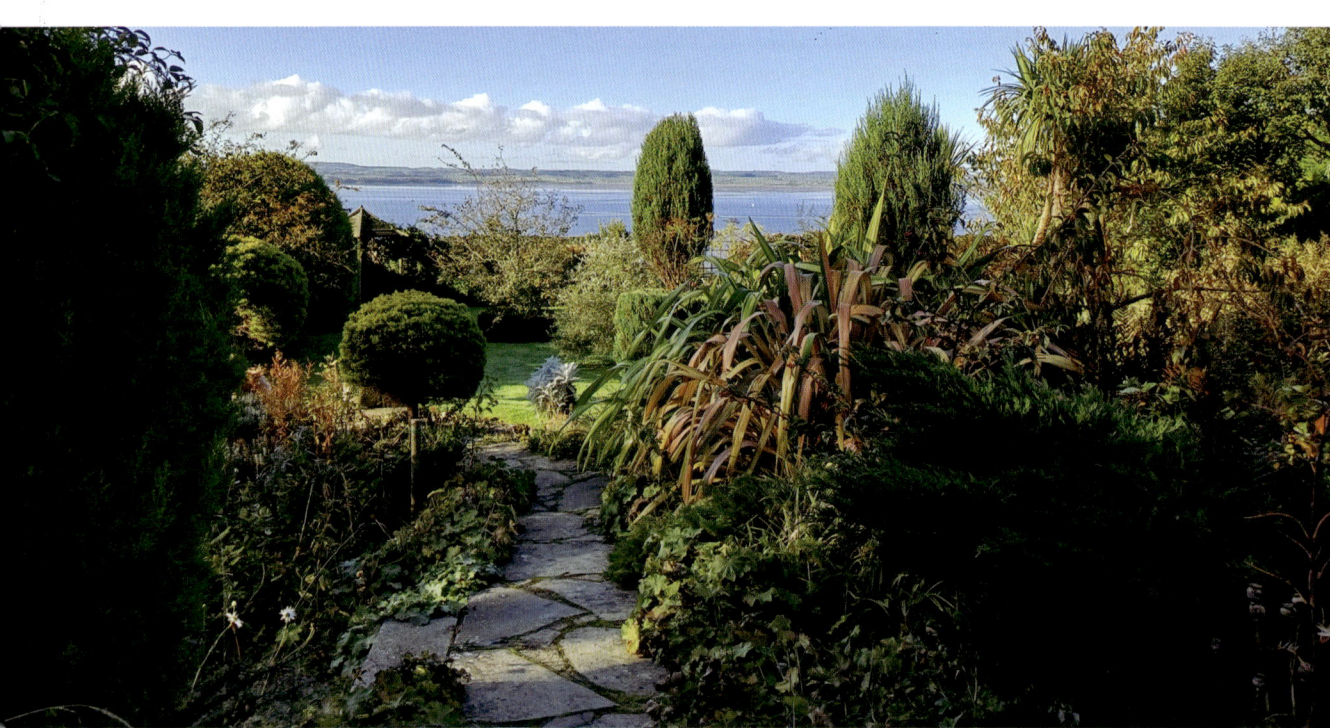

The view from the front door, showing the crazy-paving path.

The 'Himalayan' paths that we had constructed up the far side of the ravine in 2020 have given us access to the steep side of the ravine and an up-close appreciation of the mixed ancient woodland with its bluebells, daffodils, wood anemones and sorrel that clothe the slope in the spring.

Dykes

Galloway fields tend to be demarcated by drystone dykes rather than hedges. At Barholm, all of the drystone dykes needed major repairs. The walled garden walls were partially repaired by the contractors when they did the main building works, but other parts had been so encroached upon by ivy, with stems as thick as a tree trunk, that they had to be dismantled and completely rebuilt. When we bought the second part of the walled garden (see page 126) we had a great deal of repair work to do on the dykes. We inserted a gate from the walled garden into the sunken garden in a collapsed part of the dyke, with the addition of two carved stone heads that I had found in a bric-à-brac shop in Portpatrick. The gate was discovered half-buried in the ground in the new part of the walled garden.

THE DEVELOPMENT OF BARHOLM CASTLE GARDEN

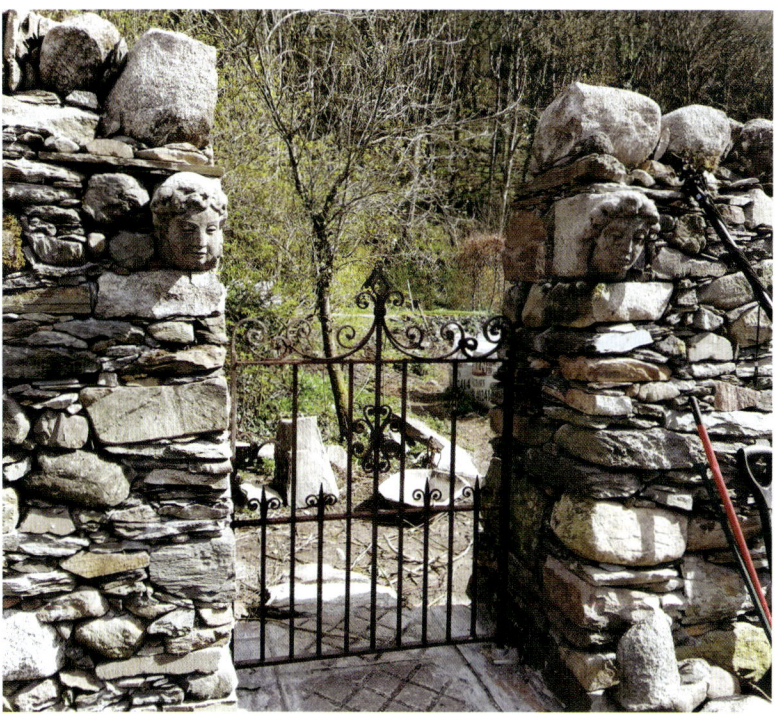 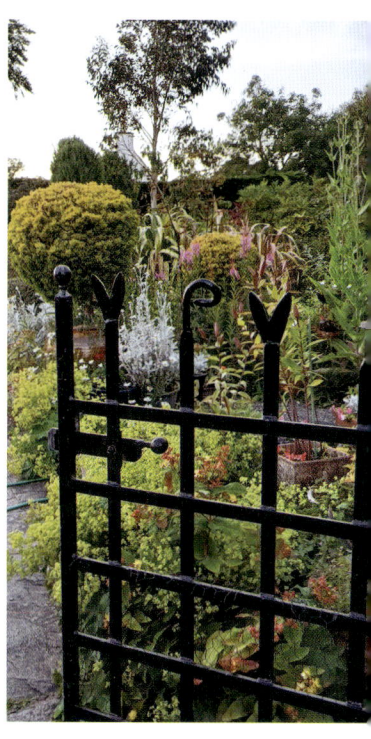

Gates

Altogether we have six gates in the garden: one standard five-bar, to keep the deer out of the paddock; two made by local blacksmith Rab Hyslop, to echo the big scarifier gate in the walled garden, giving us views over the bay; the one we dug up from the 'new bit' of the walled garden; and the pièce de resistance, the stunning thistle gate (see pages 103, 125) designed for us by artist blacksmith Adam Booth. It opens from the courtyard into the castle doorway, making a splendid entrance to Barholm Castle.

Above left.
Gate from the walled garden, with carved heads in the wall, which were inserted when repairing the drystone dyke.

Above right.
Gate into the walled garden, designed by blacksmith Rab Hyslop to echo the plough scarifier gate (see page 124).

Benches

We have twelve garden benches altogether, although I rarely have time to sit on any of them. Next door's cat, Musky, likes to do a ritualistic tour of the garden with us, stopping at each bench that we pass, so that she can sit down and have her ears scratched. Most of the benches have some special significance. The local blacksmith, Rab Hyslop, made one that echoes the shape of the 'portcullis' gate in the walled garden. Apprentices from Whithorn

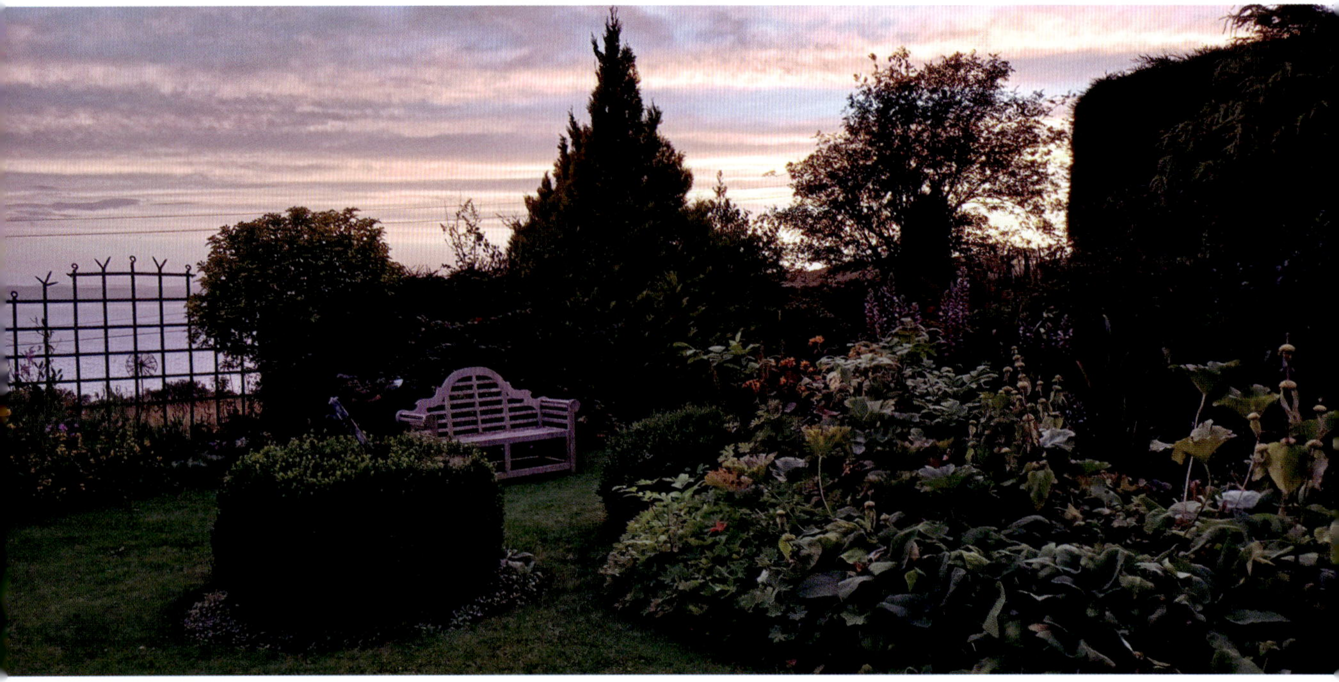

Above.
A traditional Lutyens bench in the walled garden.

Right.
Bench in the sitooterie beside a table made by Jim Walker from one of our fallen elms.

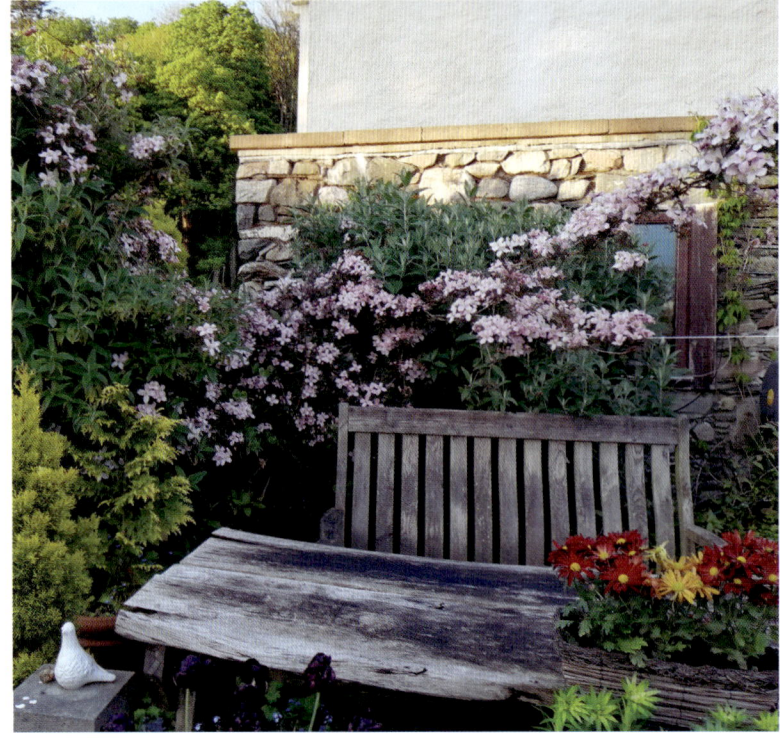

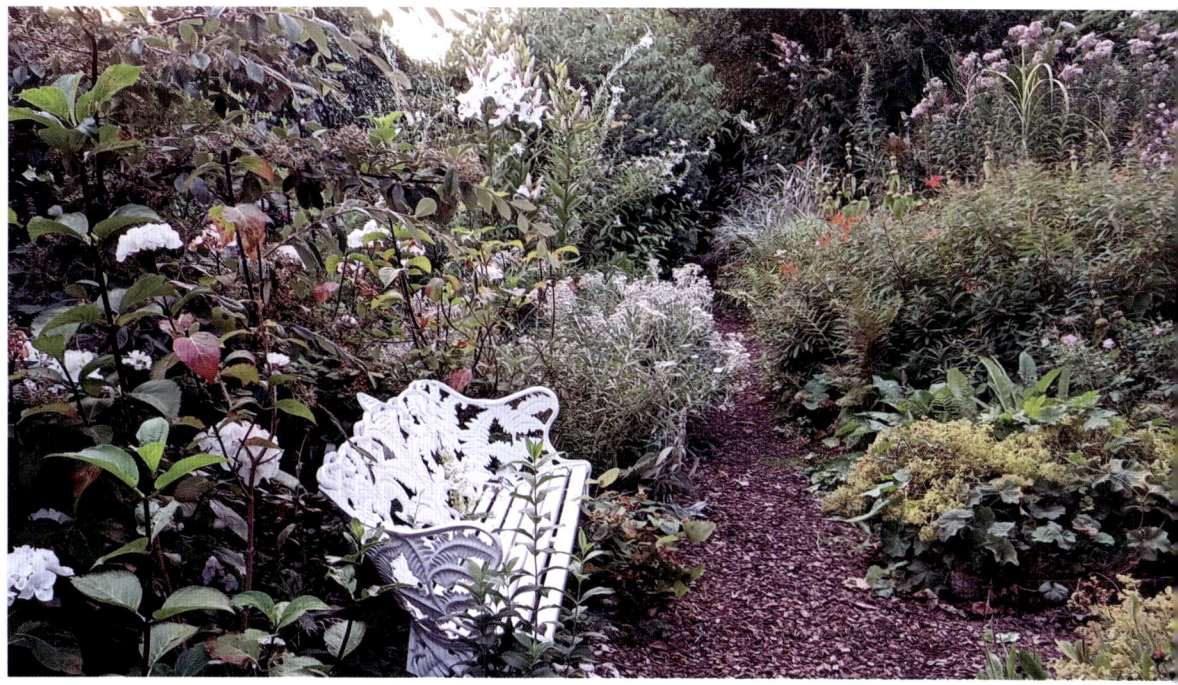

A reproduction white Coalbrookdale bench in the white border of the walled garden.

made a simple oak and elm bench with no screws or nails. It sits by the paddock pond. We have two replica Coalbrookdale fern benches, a green one in the ravine and a white one in the white border. A Lutyens bench is placed by the pond in the walled garden, giving a view of the castle. And a leaving present from friends in the Netherlands is in the north part of the walled garden, on the lawn that I sowed.

The mound

In 2011 we had a large conical mound constructed out of a heap of building spoil and we had the grass covering it mowed into a spiral with a narrow path winding round up to the top. Over the years, it became more and more of a chore to keep this piece of labour-intensive landscape art looking smart, like the pristine Garden of Cosmic Speculation (Dumfries and Galloway) which originally inspired it. Encouraged by the incursion of lots of ox-eye daisies, I decided to transform it into a wildflower mound. Establishing a long-flowering wildflower area in grass is not easy, particularly when the first clearing and sowing of seed is followed by a six-week drought, but the mound is gradually becoming colonised by a wide variety of annual wildflowers – poppies,

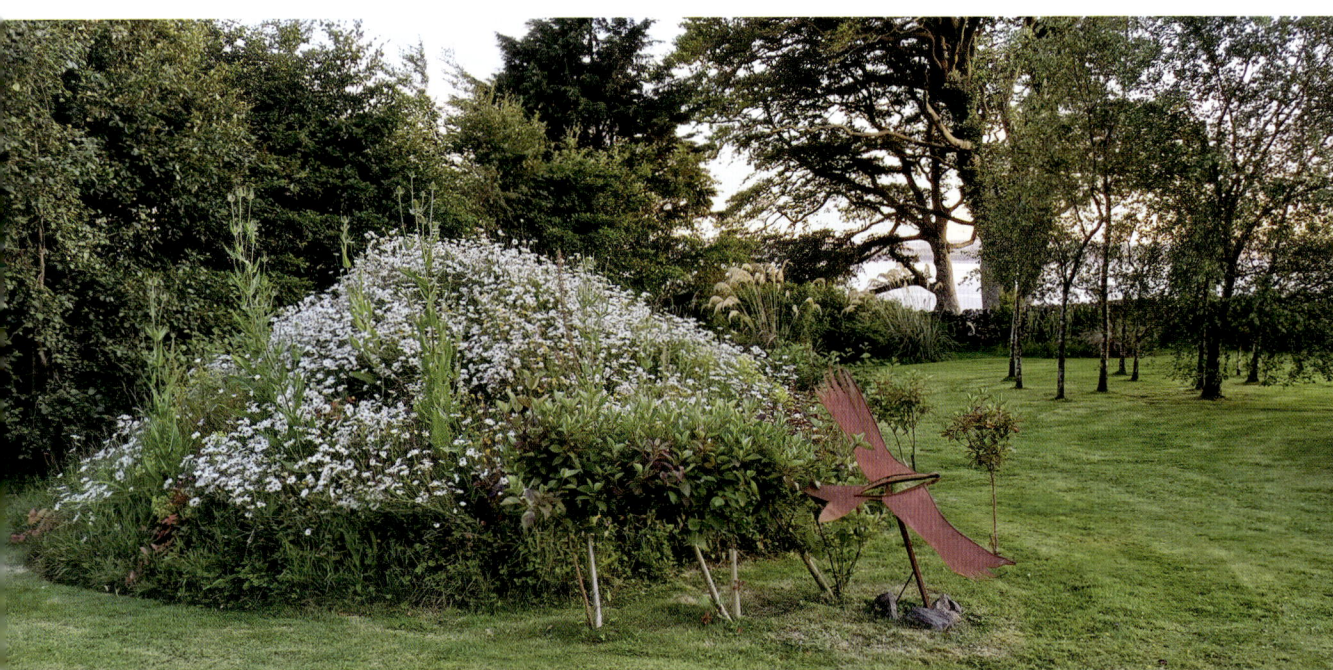

The mound today, with ox-eye daisies and teasels in full flower and a red kite sculpture in front. Red kites used to be a rare sight when we first moved to Barholm but now they fly over the garden in increasing numbers.

red campion, wild geranium, echium, bird's foot trefoil, ox-eye daisies, yarrow, etc. – and a few helpful garden perennials, such as *Anaphalis*, *Lysimachia*, *Pulmonaria* and *Alchemilla mollis*, which will settle and spread happily in a crowded environment. The job of the gardener is to adjudicate in the struggles for horticultural dominance and ensure fairness for all in the jostling for position, and that task keeps me endlessly occupied.

Sculpture

We have installed a number of pieces of sculpture, large and small, throughout the garden. The giant poppy heads (see page 104), which look as though they are made of ferrous metal, are actually carved out of red sandstone. The red kite by the mound is ferrous metal, however. A couple of Lewis chessmen have weathered nicely and almost look as though they are part of the planting, and a stone duck that we placed at the side of one of the ponds has also settled in well.

THE DEVELOPMENT OF BARHOLM CASTLE GARDEN

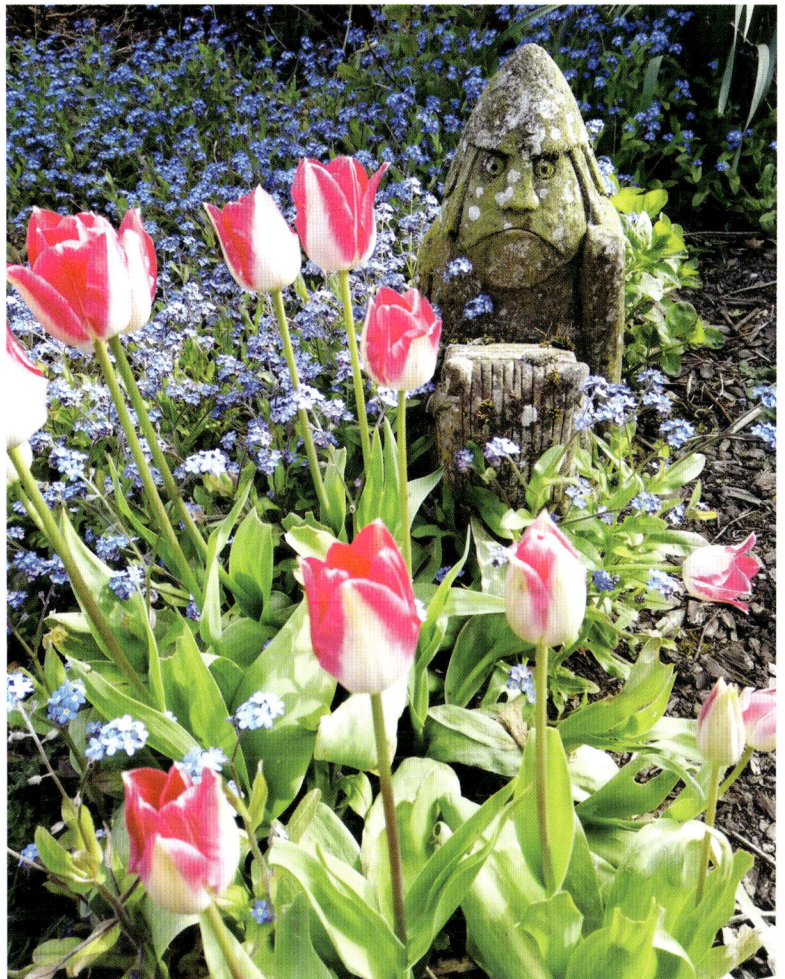

Left.
A Lewis chess knight almost hidden amongst the tulips and forget-me-nots.

Below.
Over the course of twelve years, the shiny black stone duck was colonised by lichens and mosses and now looks ancient. Ginger the cat passed away some years ago and is still sorely missed.

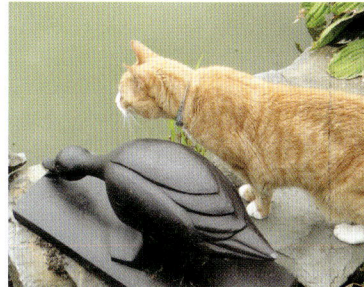

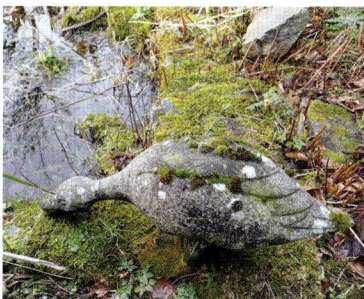

The standing stones

In the courtyard next to the castle are three eye-catching 'standing stones' which visitors assume are of great antiquity. They are correct. But although the stones are truly ancient, the reason they are in the courtyard was to hold up the roof of an eighteenth-century farm byre which abutted the castle's north wall. The *Second Statistical Account* of 1840 describes how local farmers used to make off with stones from Cairn Holy, the 5,000-year-old Neolithic chambered cairn burial site near us, and use them for building projects. Brian Blake, author of *The Solway Firth*, visited Cairn Holy: 'The two great tombs up here tell any amateur that they are the remains of once great covered chambered cairns, and when I saw them I knew at once where the long stones at Barholm had come from' (1966, p. 176). To be fair to the farmers, before the excavations at Cairn Holy in 1949 and the erection of interpretation boards by Historic Scotland in 1957, the sites were not nearly so obviously demarcated as tombs. The HES Statement of Significance states that both tombs were substantially robbed of their stone for use in building nearby field dykes in the eighteenth century. They were robbed for more than just field dykes, it seems.

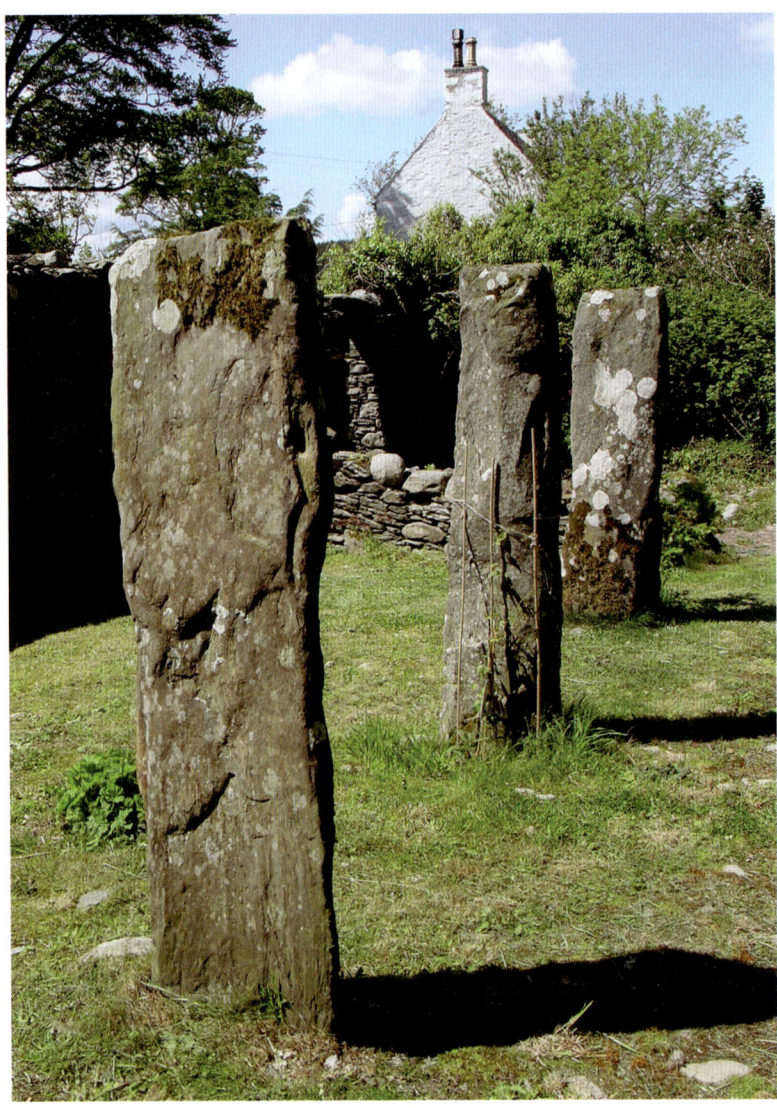

The 'standing stones' in the courtyard, almost certainly robbed from Cairn Holy in the eighteenth century and used to support the roof of a byre. I grow clematis and honeysuckle against them.

The courtyard is cobbled, as it was used as part of the farmyard. There are a few low-growing plants between the cobbles and the 'standing stones' in the middle and some shrubs against the wall. The main plant features are in two low stone walls with planting pockets and a long outcrop of rock which floods in wet weather and in which some primroses are very happy. Clematis and honeysuckle are trained over one of the standing stones.

Hedges and trees

Plants can be structures too, of course. We have planted *Cornus alba* (Siberian dogwood) and beech as hedges in the paddock. Along the border with the neighbours in Barholm House we planted a long wildlife hedge, mostly of native trees and shrubs. We have let it run wild, rather than hedging it every year, and it now provides food and shelter for dozens of species. In the walled garden, the very first plants I put in were six box plants to start off a little hedge. That hedge is thick and topiarised into waves now, but part of it was destroyed by blight a few years ago, and what is left has been badly attacked and sadly may have to be grubbed up. We have a thick mature ivy hedge by the track; butterflies feed on the flower nectar in autumn and birds eat the nutritious black berries. I would like to cloud topiarise the hedge, but it is growing into the top of the ravine and I would need skyhooks to hold me in place for the operation. The 15-foot high *Cupressus leylandii* hedge that separates the two halves of the walled garden is also a haven for wildlife. Wood pigeons are particularly fond of hiding and nesting in it; they make a noisy clatter of wings when disturbed, startling both of us.

From the very start, I have planted trees – hundreds of them – throughout the paddock and walled garden to provide structure and interest, and as a wildlife habitat. If you are starting a large garden, trees and shrubs should come first. They generally take a long time to grow, so get them in early. One of the first trees I planted was a wild rowan seedling, a gift from the owners of A.D. Livingston and Sons in Castle Douglas, who provided our oak kitchen table many years ago. The rowan is now a mature tree, covered in lichen – a sign of good air quality. The orange berries never last long in autumn, as the birds strip them quickly. There is a little glade of silver birches in the paddock which is finally casting dappled shade. Some of the less common trees now thriving in the garden are *Nothofagus antarctica, Metasequoia glyptostroboides* 'gold rush', *Cercidiphyllum japonicum* (the toffee apple or katsura tree), *Cedar deodora, Cryptomeria japonica, Liquidamber, Liriodendron, Embothrium lanceolatum* (the Chilean fire bush), *Crinodendron* and *Eucryphia*. The last three have spectacular displays of unusual flowers.

I have planted more eucalyptus than any other tree, including four different varieties. They grow very quickly, which is a great advantage when establishing a large garden from scratch, and their silvery leaves and attractive bark provide a good foil for some of the brighter and darker green shrubs in the garden. Bees love the blossom too. Some of the trees that I have planted are non-native, but many are either naturalised or native: the alders, *Amelanchiers*, ash, beeches, birches, box, dogwoods, elders, holly, laburnums, larch, privet, rowans, yew, field maples, spindles, willows and wild cherries. We also have a number of oaks on the wooded slope above the ravine and a large ancient oak by our entrance. Such a tree can support over 2,000 other species, including birds, bats and insects – and, of course, the red squirrels that we see scampering up and down its branches.

Pruning and topiary

In the walled garden and the paddock most of the conifer and evergreen trees that I planted more than ten years ago have now become mature enough to be shaped. *Cupressus macrocarpa* 'Goldcrest', which has a sweet lemon scent and a bright golden lime-green colour, is easy to trim into sharp shapes, and we have a dozen or so. Bay trees and hollies have finally grown enough to be shaped into balls and cones, and a yew that was given to me by my aunt and uncle when we completed the restoration is now a sturdy pyramid much taller than me. All of these trees, shrubs and hedges give structure to the garden, so that it is never looks dull or 'dead' in the winter. On the few days each year when we have snow, the topiary looks especially striking.

Pruning is something I am not keen on. I know that the roses will flower better for it, the cornus will have a greater depth of winter colour and the buddleja will have larger flowers lower down its branches, so I do prune these. Some rampant growers need to be cut back to allow access to paths. But, on the whole, I tend to trim lightly and leave most shrubs to get on with growing the way they wish. In spring, the flowering currants and forsythia make magnificent statements, as they are rarely cut back and have become huge where there is plenty of room for them. If shrubs do get out of hand and fill up far more than their allotted space, then sadly they have to go. In 2009 I planted two little *Fuchsia 'Riccartonii'* each side of the steps in the top of the retaining terrace wall in the walled garden. In 2023, having reached a height

THE DEVELOPMENT OF BARHOLM CASTLE GARDEN

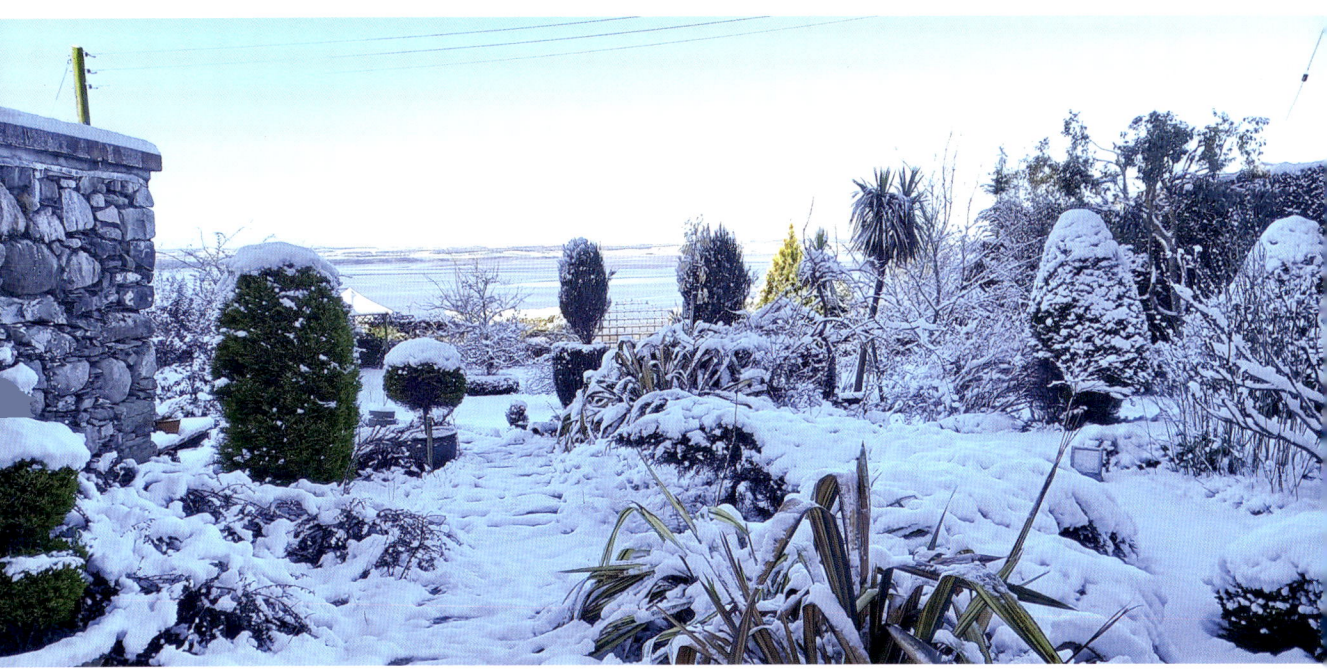

The garden on a rare day of snow, with the topiary shrubs coated like Christmas puddings.

and spread of almost ten feet, they had to be removed. The harder you prune a fast-growing shrub, the more vigorously it will grow back. I was tired of having these hacked down every spring only to watch them grow another 6 feet in all directions. They have left a welcome legacy of two large open spaces to be filled with new plants, including a couple of little fuchsia seedlings that have sprung up and will be left as long as they remain at a reasonable size. The giant yellow tree peony in the new part of the walled garden had got out of hand; when I had it cut down, hundreds of little seedlings flourished in its place, ready for transplanting for future plant sales.

The ponds

We put in three wildlife ponds very early on: one in the walled garden, one in the paddock and one in the ravine. Word clearly got around the local amphibians very quickly, as each spring we found clumps and strings of spawn in the ponds, developing into a cast of frogs, toads and newts – a case of 'if you build it, they will come'. The first spawn is usually deposited by frogs in February and almost inevitably followed soon afterwards by a period of freezing weather, when ice forms over the clumps. We then worry about

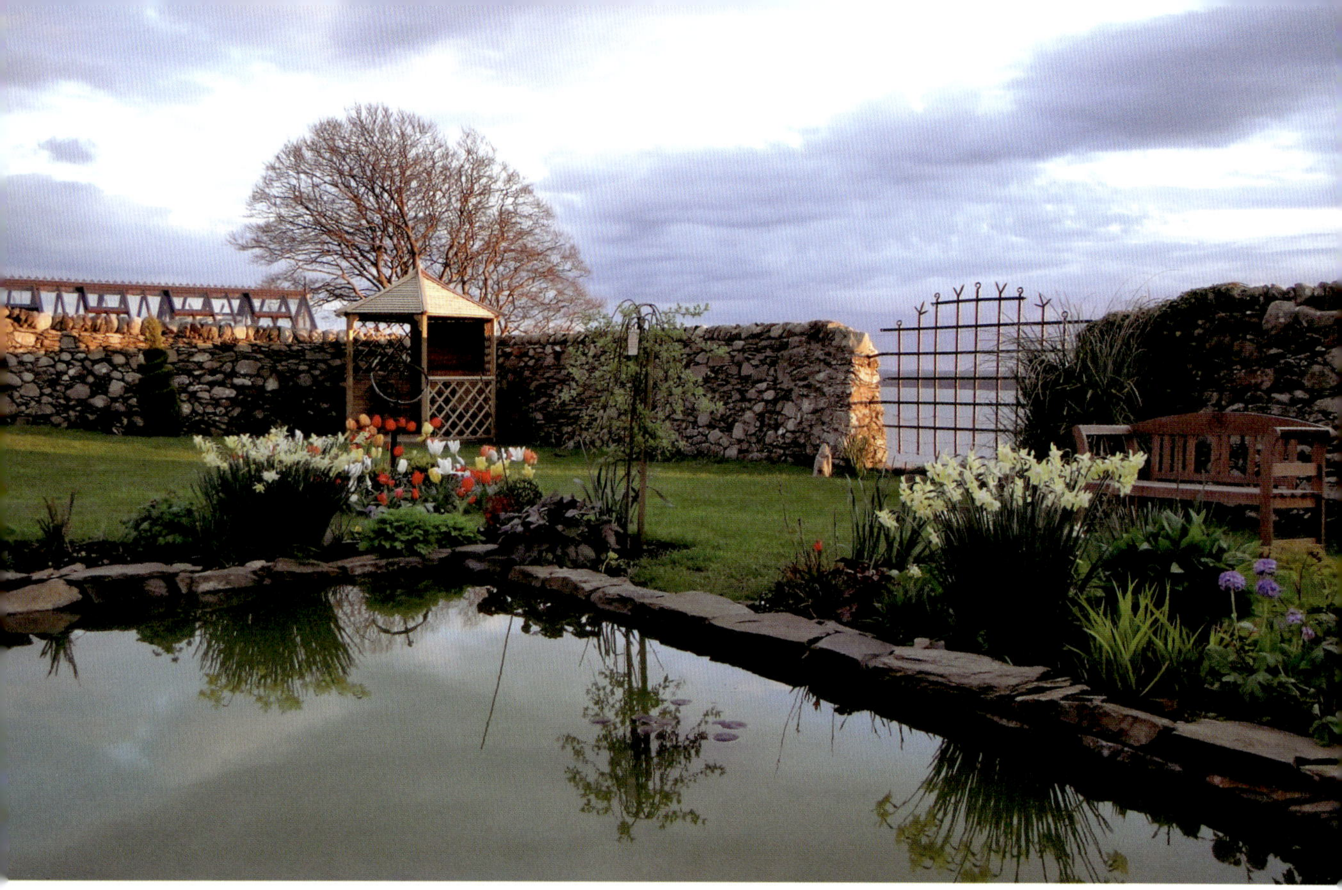

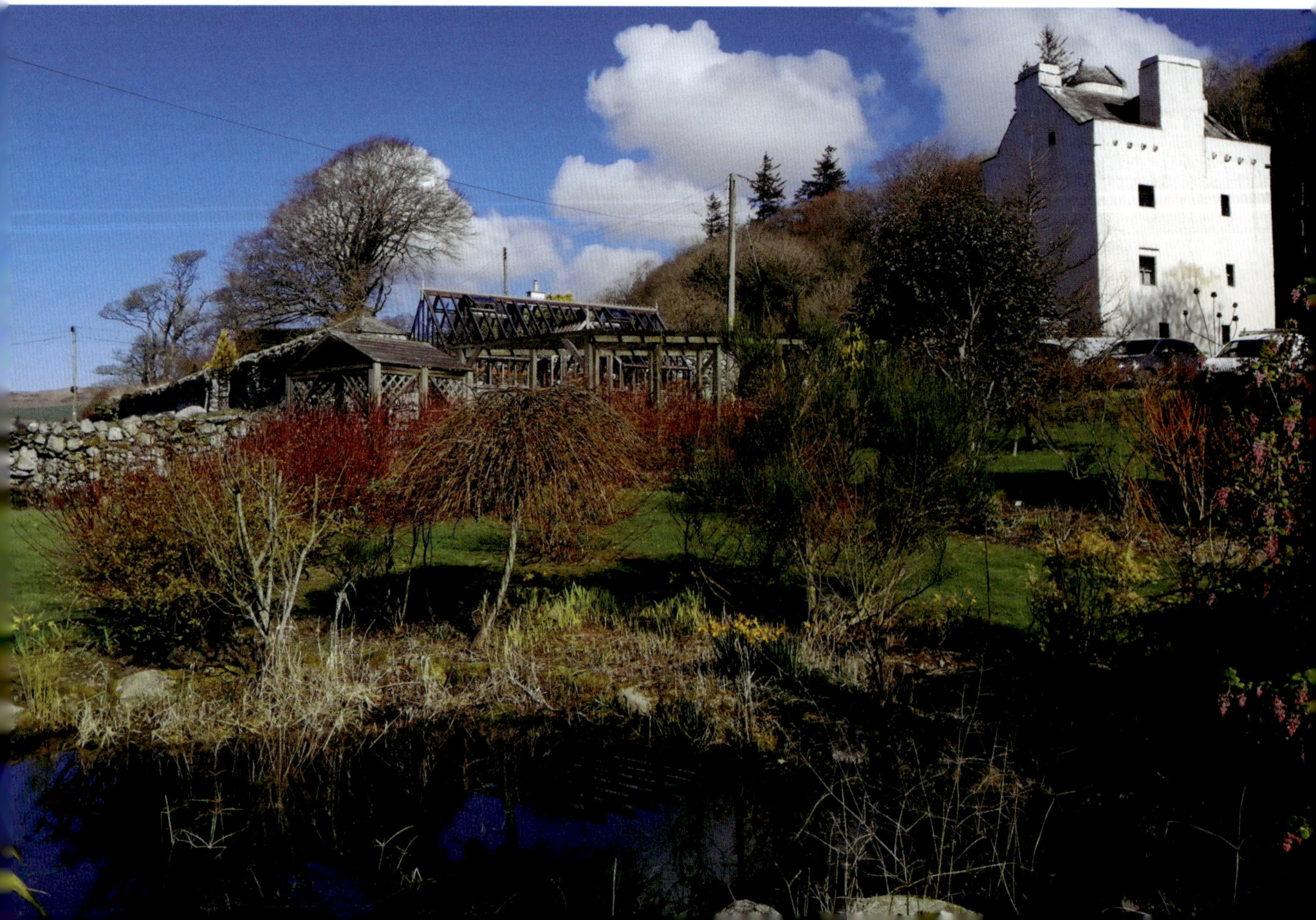

the viability of the spawn, but every year, somehow or other, they survive to become tadpoles and froglets. Around all three ponds we have planted grasses, shrubs and perennials to attract birds and insects, and in the ponds a number of aquatic plants, including different species of water lilies, to provide shade and shelter for the occupants.

The sitooterie (where we sit oot)

Garden design is usually constrained by the physical features of the site. At Barholm we were advised to plant nothing in the area beside the boiler house, because the large copper earth beds of the lightning conductors lie underneath. We started with the best of intentions, as can be seen from the 'before' photograph, showing clean pea gravel topped with nothing more than a table and chairs – a sitooterie. Now it is a different picture, although the area of the lightning conductor beds is more or less clear of planting under the table, and much of the planting is in movable tubs. It is still a great spot for sitting out, but now surrounded by plants, where we occasionally enjoy a summer glass of prosecco, looking out at the wonderful views across the bay.

Opposite top.
The walled garden pond in the evening sun.

Opposite bottom.
The paddock pond in spring.

The sitooterie before I started planting.

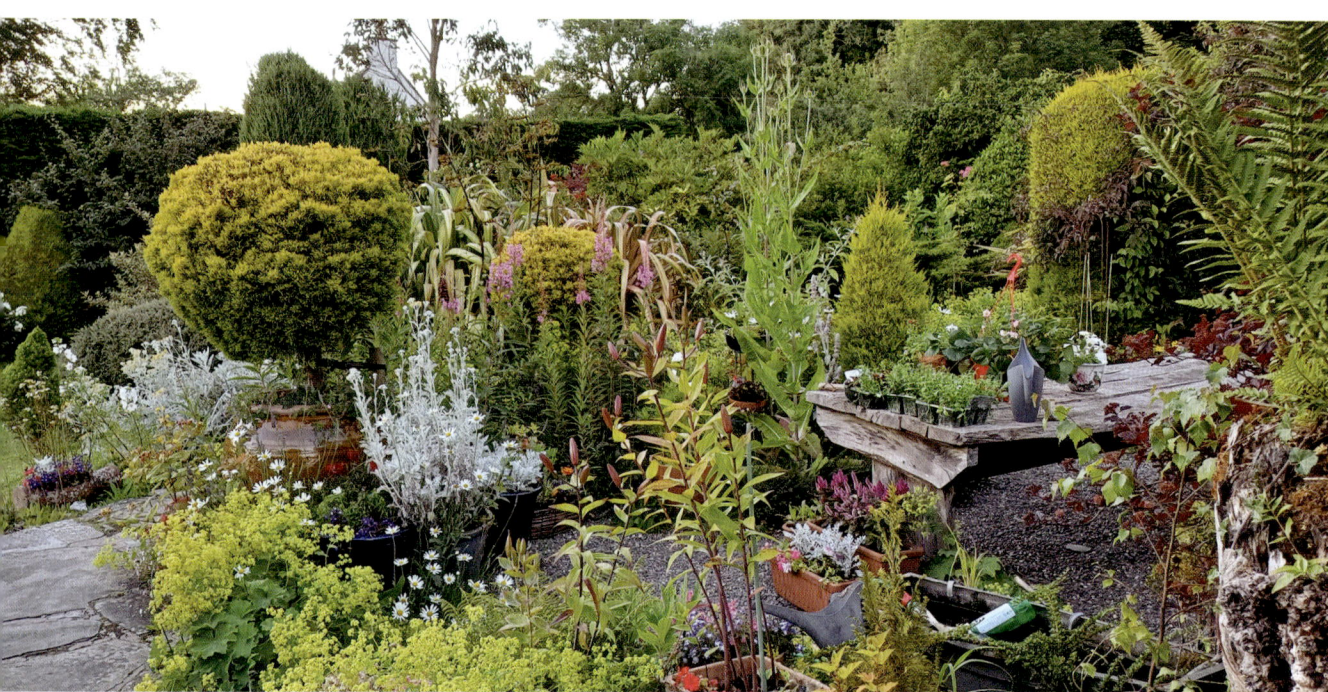

Island beds

The sitooterie today.

Island beds are a concept of the 1970s, made fashionable by horticulturalist Alan Bloom at Bressingham Gardens, Norfolk. The idea was to make beds that look good from every angle, filled mainly with perennials. Island beds are ideal for large gardens and I have made four in the walled garden and three in the paddock garden. The first island bed was made in the walled garden because a pile of builders' material had been stacked there for a couple of years and when it was moved the ground was weed-free in an oval patch. My late cousin Christine came with a truck load of her unwanted shrubs and we dug them in hastily in a rainstorm. Some did not survive, but the mock orange and golden *Cupressus* are still thriving, now transplanted elsewhere. The paddock garden island beds took several years to look good – the poor soil, high winds and predation by rabbits militated against success – but finally they look established and break up the large space nicely.

THE DEVELOPMENT OF BARHOLM CASTLE GARDEN

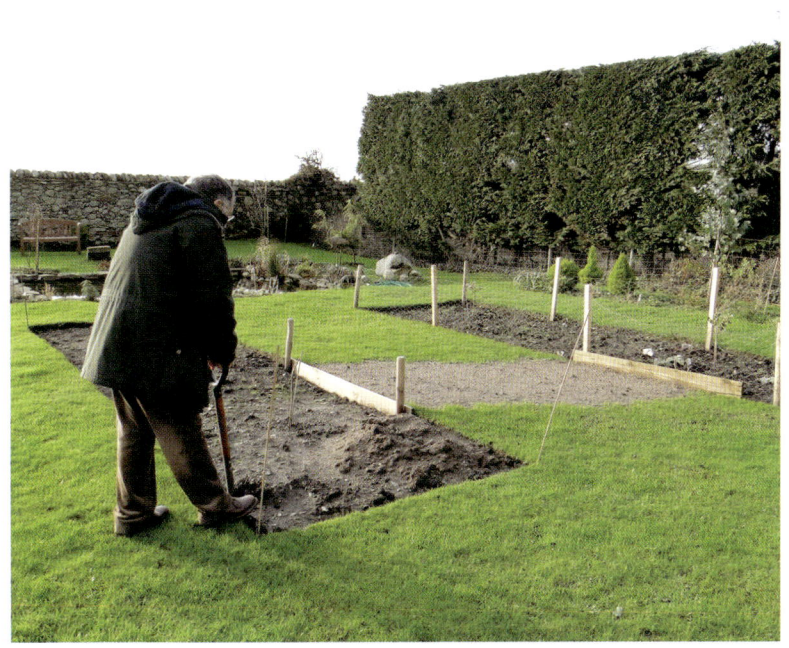

Left.
The first island beds in the walled garden, ready for planting in early 2012.

Below.
The same island beds in 2023.

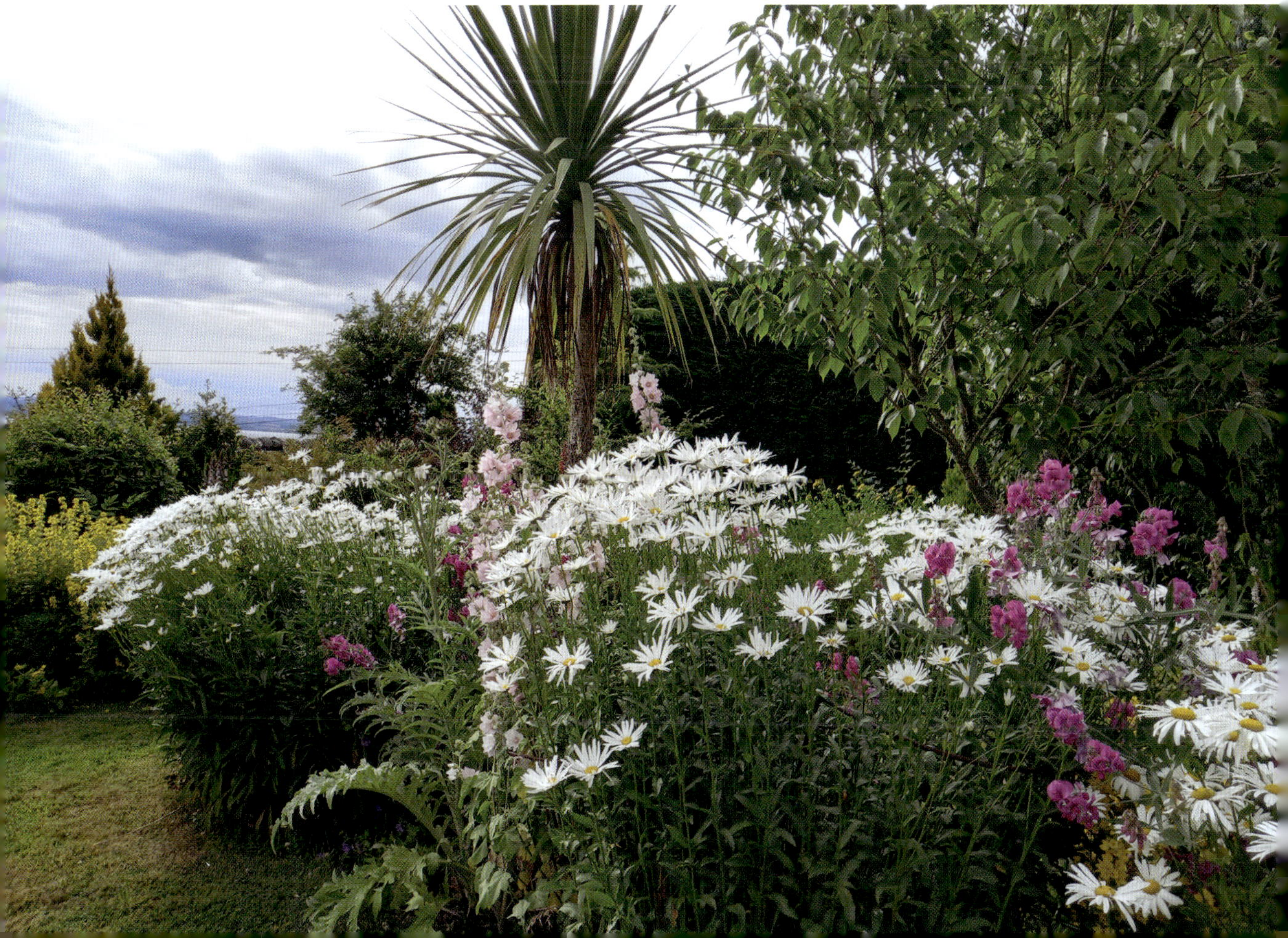

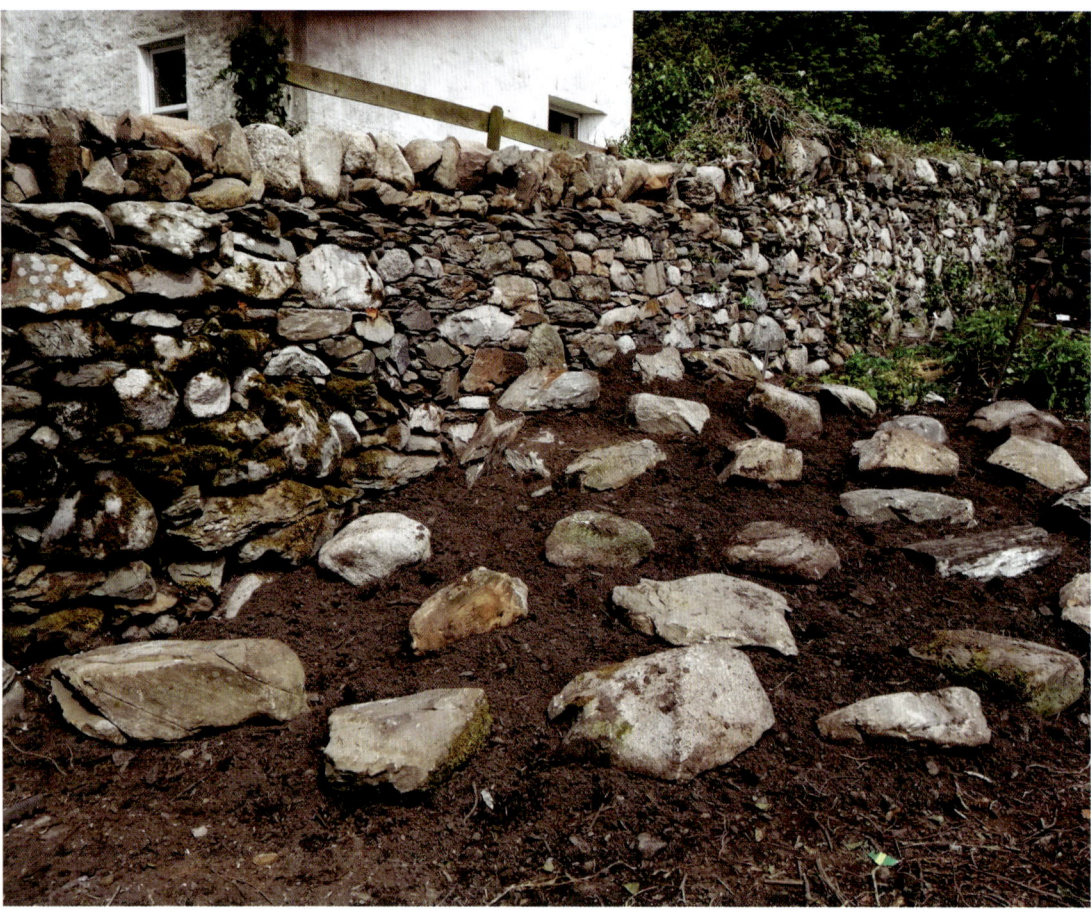

The walled garden rockery newly constructed, before being planted.

Rockeries

With so much stone lying around and so many changes in levels, parts of the garden seemed to be crying out for rockeries. At the last count I had seven. Not the carefully managed crevice gardens that might earn me the admiration of alpine enthusiasts, but sloping, terraced areas of very large stones planted up with blousy, billowy garden flowers. *Nepeta* 'Six Hills Giant' looks wonderful trailing down across stones, and well-timed cutting back after the first flush of flowers gives the opportunity for a second flush in autumn. Fuchsias, persicarias, periwinkles and potentillas provide structure, and drama comes from the occasional statement cardoon or giant *Echium* or *Crambe*.

In the new part of the walled garden there was a set of ugly concrete steps

THE DEVELOPMENT OF BARHOLM CASTLE GARDEN

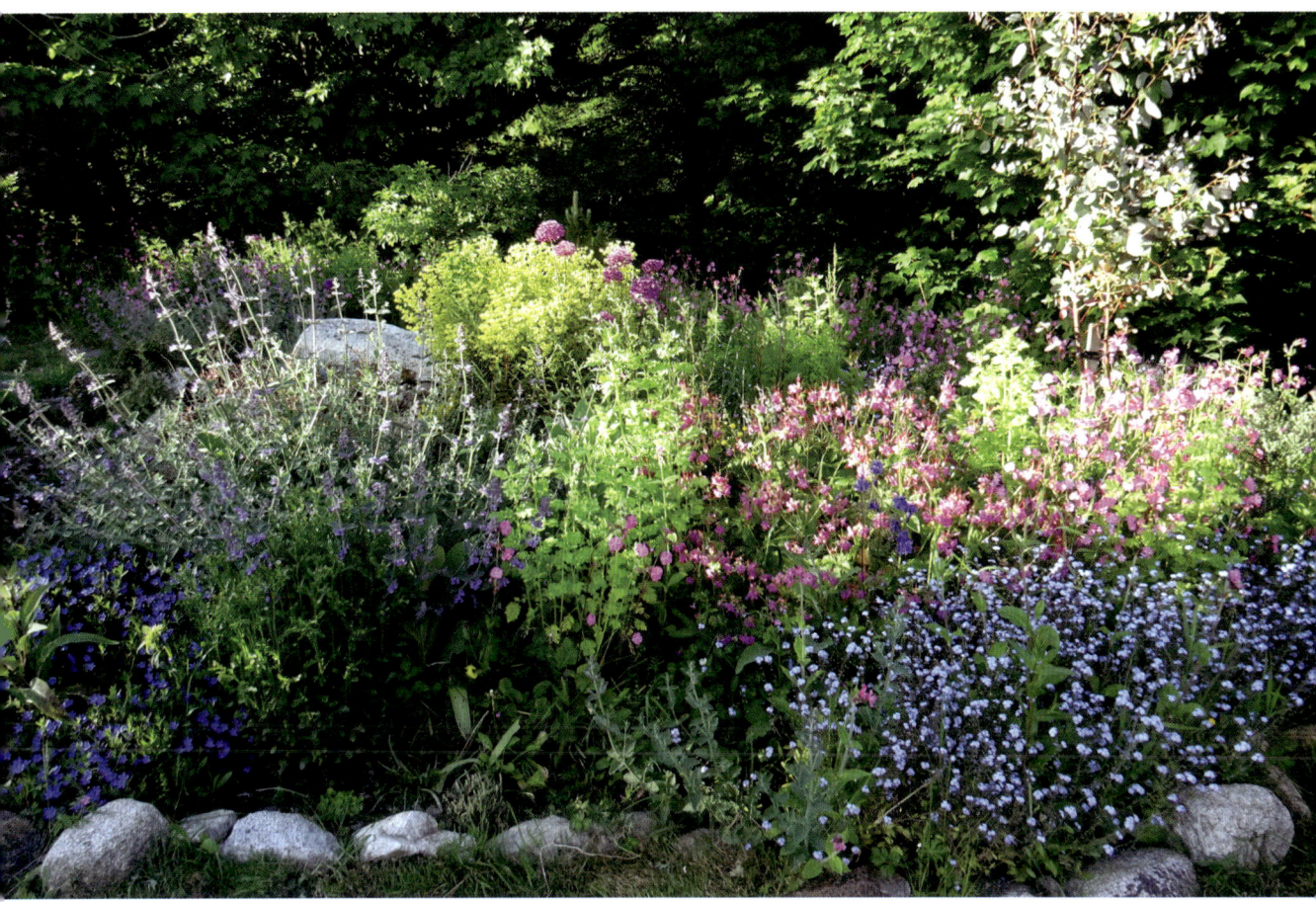

leading down from the farmhouse garden, now blocked off. We could see no easy way to get rid of the steps, so we covered them in soil and large stones and planted out a rockery, which established remarkably quickly. The change in levels in the sunken garden has given opportunities for a couple of small rockeries beside the steps we installed; these are filled with spring bulbs (with anemone de Caen making a much-admired splash) and rockery perennials.

During lockdown, John constructed a fern rockery in the ravine from stones that had fallen from the retaining dyke and become encrusted with mosses. In it he planted Japanese painted ferns, royal ferns, tatting ferns, Welsh ferns, button ferns and several others. It is now an attractive feature beside the path on the way down to the main fern garden.

The rockery by the road, made from a heap of builders' rubble and brambles.

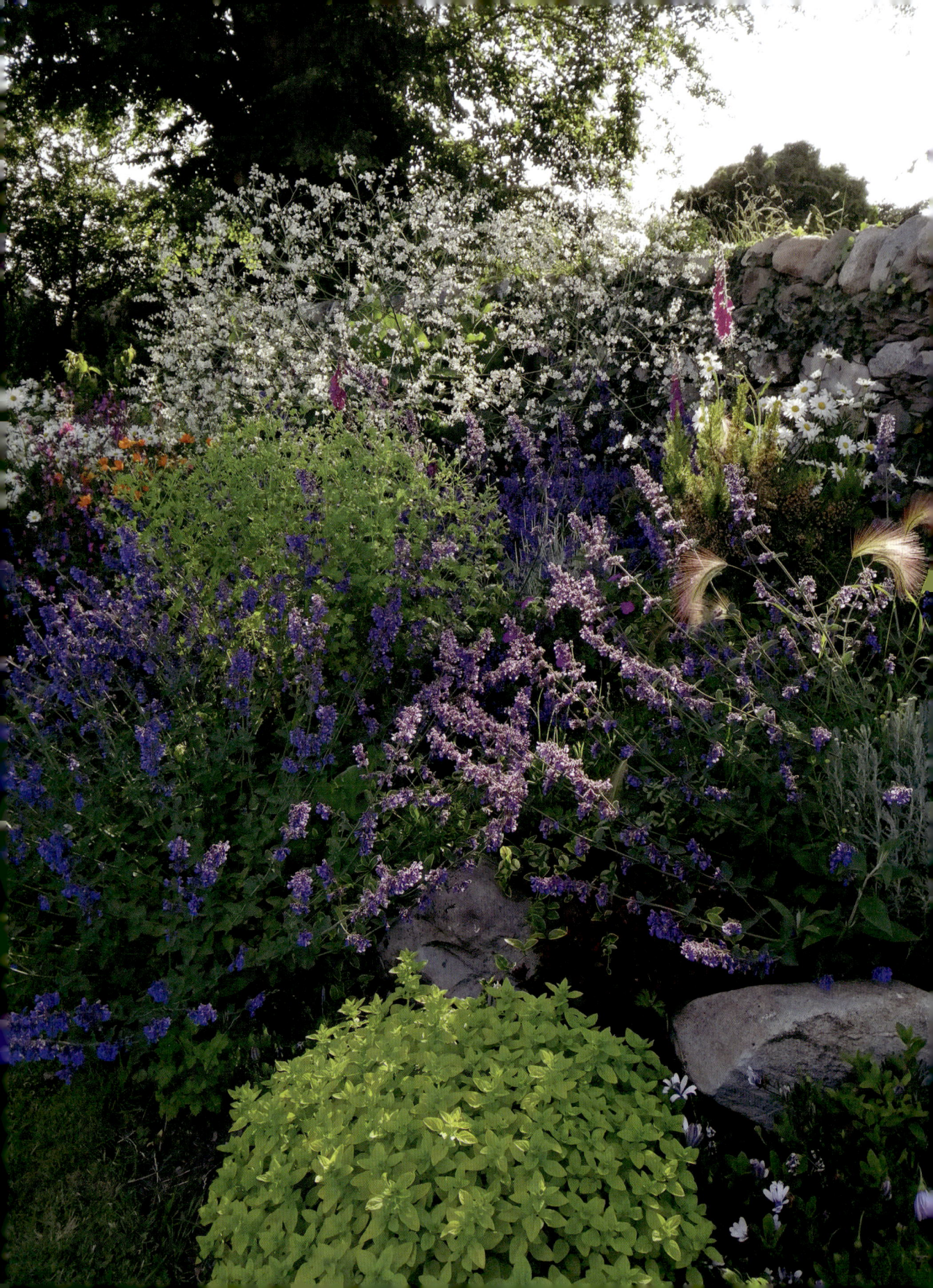

Reflections on Gardening

Gardening today

It seems to me that we have come to a crisis, or at least a turning point, in our understanding of what a garden is and what it is for. I recently picked up in a charity shop a copy of a BBC '*Gardeners' Question Time*' book from the 1970s and was surprised to see how often the answers to questions – which were nearly all preoccupied with pests and diseases – contained advice to use chemical pesticides, weedkillers and/or fungicides. Killing something was the standard answer to every problem. The indiscriminate use of toxic sodium chlorate, outlawed since 2009, was routinely used to kill weeds; the days of glyphosate, although an altogether safer compound, are now seriously numbered. If future gardeners want to get rid of weeds in their gardens, they will have to pull, dig or hoe them out. Thankfully, we have become much more tolerant of 'untidiness' in gardens and there is a laissez-faire attitude towards self-seeders, many of which sustain wildlife. The use of power tools is increasingly frowned upon for environmental reasons, particularly those that use petrol. Jim uses quiet lithium battery tools when cutting hedges and mowing the grass in the walled garden, although we still use a tractor mower once a fortnight in the paddock.

All gardening is unnatural, as we try to impose our idea of aesthetic order on nature, but the emphasis now, quite rightly, is on sustainability, biodiversity and the promotion of plants that are attractive to pollinators. The threat of climate change has driven changes in our gardening habits, along with the realisation that importing plants from overseas can lead to the introduction of disease. Brexit has resulted in cutting off the UK from many of the varied garden plant imports from the European Union. The days of British gardens as neatly groomed show-off repositories for exotic plants from overseas are necessarily numbered. But we can still be gloriously colourful gardeners. In the west coast of Scotland we have the double benefits of high rainfall and a temperate climate, and here at Barholm we have the additional assistance of some shelter from the winds and very fertile soil in the walled garden and the ravine. Gardening will go on, even if its practice changes.

Opposite.
The planted rockery hides a set of ugly concrete steps in the walled garden.

Garden snobbery and fashion

There is a breed of 'haughtyculturalists' who police the plants we grow in our gardens. Theirs is a tough job, saving us all from hideous horticultural faux pas, because the list of undesirables is constantly changing. I noticed some very large and superbly lush clumps of pampas grass in King Charles's garden at Dumfries House recently and wondered how long it will be before the former pampas haters realise that it is now officially regally fashionable once again. Pampas grass has been infra dig for years, although, strangely, those same people who sneered at pampas grass became enchanted by less-available *Calamagrostis* which is quite similar.

In the hierarchy of horticultural styles, plant favourites from two generations ago fare worst. However, after fifty years in the outer circle of poor taste, these plants will suddenly be rediscovered and pushed forward as the best thing ever to happen to gardens. Dahlias, for so long regarded with contempt, have recently become extremely fashionable, as have lupins. Begonias are still beyond the pale, but I predict that they are due for rehabilitation. Likewise hydrangeas, which, according to gardening writer Christopher Lloyd 'are regarded, in refined circles, as crude, blatant, obvious, coarse, vulgar', to which he added 'In that case, I must have something of all those qualities myself' (quoted in Foster 2023, p. 15). Hydrangeas are my very

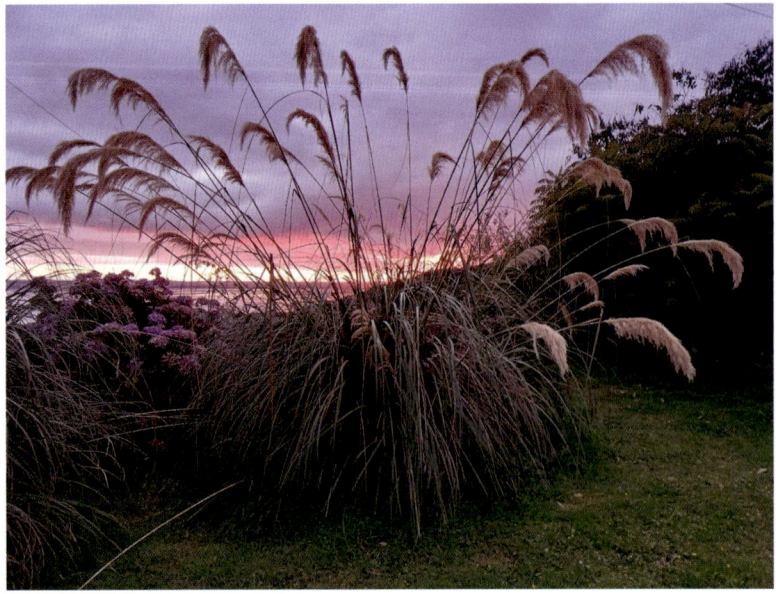

Pampas grasses filter the winds, acting as a kind of transparent shelter belt, and look fantastic against the sunset.

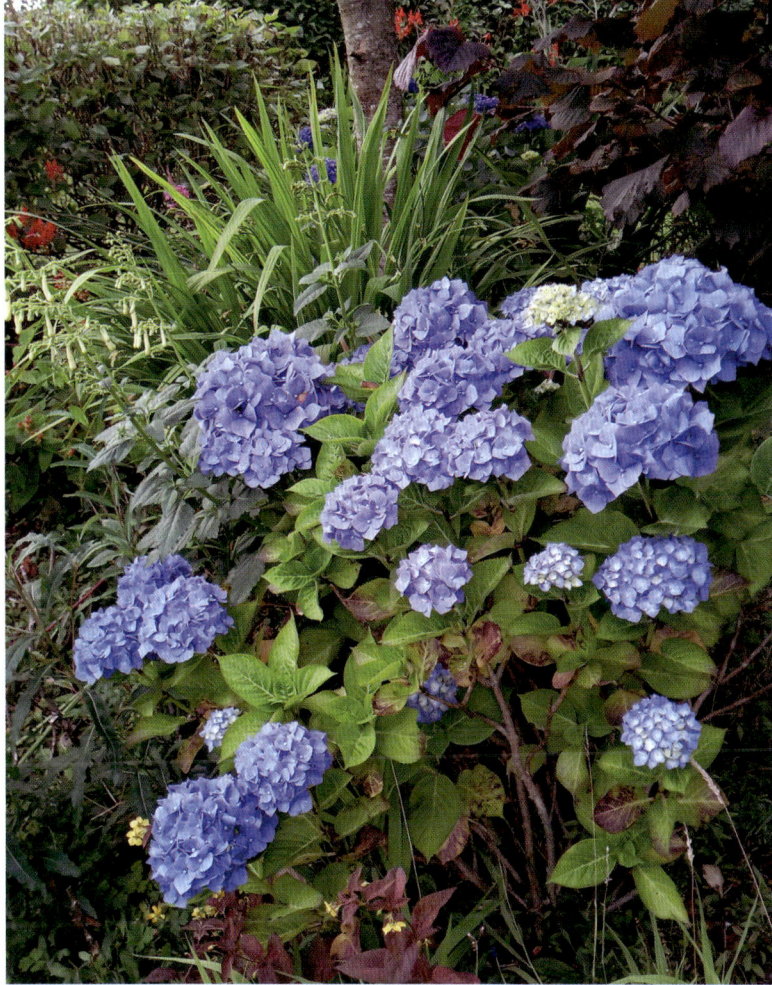

The vivid blues of the hydrangeas in the paddock island beds, well away from the colour-changing lime of the walls and dykes.

favourite shrubs. They flower robustly from early summer to Christmas in a variety of shades of blue, purple and pink, and the blue ones dry very nicely for winter decorations. We must have at least twenty-five different varieties in the garden – without them, it would be a poorer place. We could even legitimately rename Barholm as 'Hydrangea House' to reflect their status in the garden. John Ruskin, that Victorian arbiter of good taste, wrote gloomily of garden plants 'corrupted by evil communication into speckled and inharmonious colours' (Ruskin 1873, p. 185); much of what flowers in my garden would have plunged him into despair.

I sometimes listen to visitors who are concerned about my fast-growing plants such as *Persicaria*, eucalyptus and rosebay willow herb. However, I

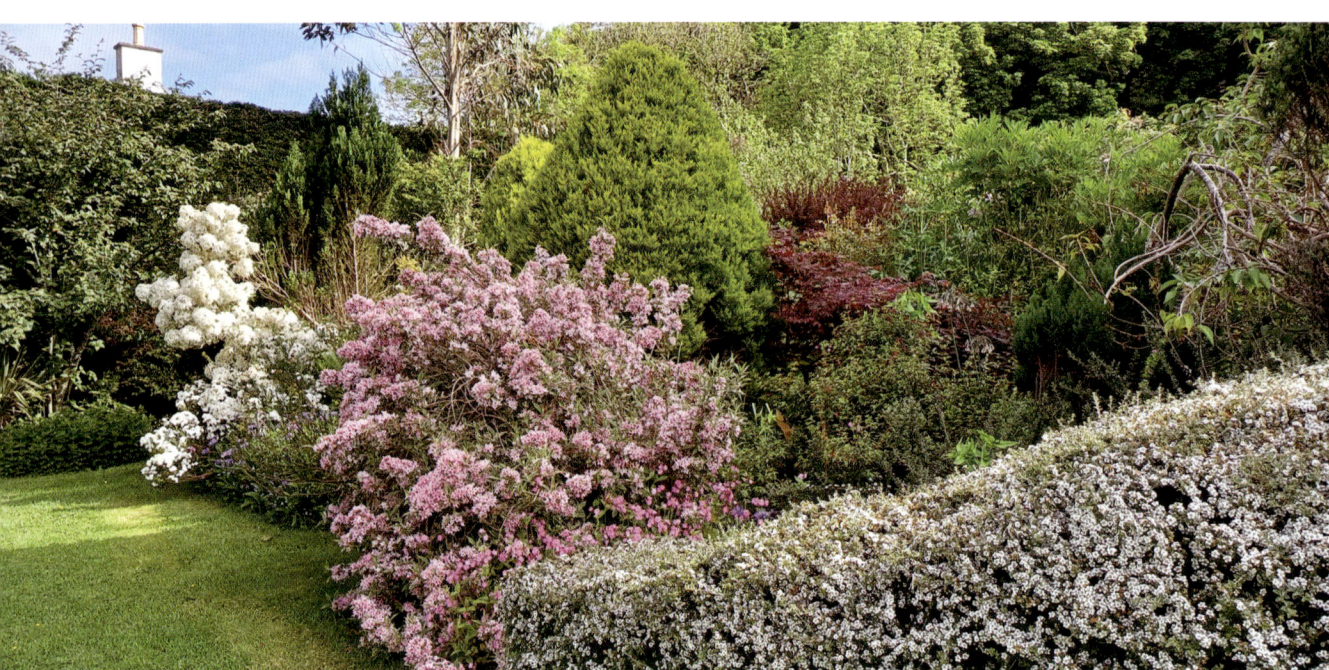

Cotoneaster microphyllus hedge in the walled garden in late spring, flowering profusely; the pink blossom behind it is a weigela and the white is an *Olearia*.

have plenty of space to let them spread, so I carry on with my planting. More often, though, visitors contribute to the knowledge I need to manage the garden, sharing their experiences and offering wise counsel. I try to follow sensible advice when gardening, but in practical terms it is simply not always possible to manage things the way they ought to be done or to do things at the optimum time. The late Christopher Lloyd, when asked at what time something or other should be done, replied 'When you can'. I hang on to that little phrase when I don't manage to get the pruning or planting done at the 'right' time. I just do it when I can.

Cotoneasters have also become unfashionable, but they are another of my favourite shrubs. I have about fifteen different varieties, from the tree-like form of *Cotoneaster watereri*, which self-seeds so prolifically that I often have to weed it out, to the miniature *Cotoneaster dammeri*. I have made a thick curving hedge out of a *Cotoneaster microphyllus*, which provides attractive shelter for acers and delicate perennials. Bees and other pollinators love the flowers, and blackbirds and thrushes strip the berries in late autumn after the frosts set in. The seeds have high viability and that is why they spread so effectively. However, I note that *Cotoneaster horizontalis* has been listed as an invasive species. It is still allowed to be grown in domestic gardens, but it is an offence to allow it to 'escape' into the wild.

THE DEVELOPMENT OF BARHOLM CASTLE GARDEN

The Seasons in the Garden

Barholm is a garden planned and planted for all four seasons. Each is a delight in its own way.

Spring

Spring is such an exciting time in the garden. The winter snowdrops have started to die away just before spring gets started with the emergence of the daffodils. There are so many different varieties, and the show goes on from February until May. The biggest display comes from those already planted beside the track before we came to Barholm. We have added thousands more. My favourite is *Narcissus* 'Tête-a-tete', the miniature narcissus with multi heads; it just keeps on spreading and is never beaten down by the wind. *Narcissus* 'Jetfire' runs a close second, with similar properties. I have planted dozens of orange *Fritillaria imperialis*, the spectacular crown imperial lily with its weird, foxy scent, but they don't come back reliably every year. Alongside the vibrant oranges, yellows and reds of the tulips and daffodils, spring is a good time of year for the complementary blues – forget-me-nots,

Spring in the walled garden.

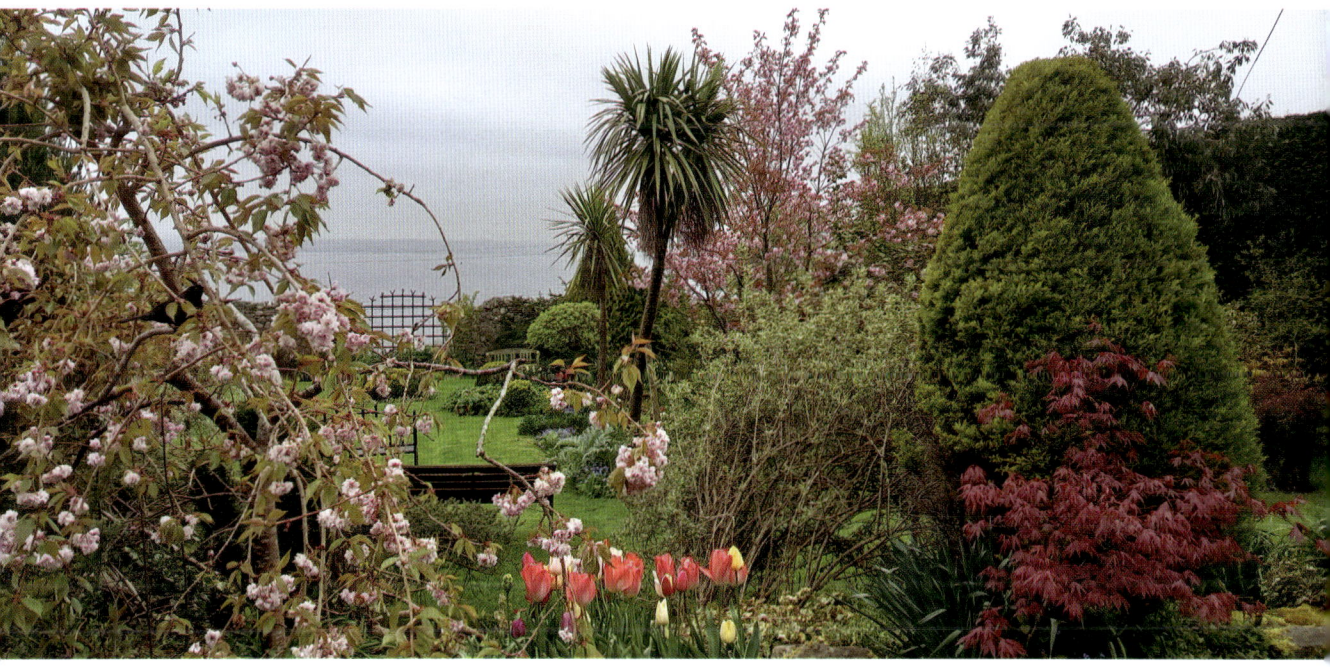

pulmonaria, grape hyacinths, scillas, *Chionodoxa* and *Anemone blanda*, followed by bluebells in late spring. I let forget-me-nots and pulmonaria self-seed throughout the garden, along with pink campion.

In the shrubbery, the large mixed Benmore hybrid rhododendrons, camellias and lilac are in flower and the new growth on the great pieris by the road is a shining red. Two old favourites, forsythia and flowering currant, also make a great splash of colour. The courtyard walls and the castle rock are full of colour provided by rockery perennials. In late spring, bluebells are dotted through the woods, just as the croziers of the ferns are starting to unfurl. Every year they seem to appear earlier than the year before.

Summer

In summer the walled garden and the paddock glow with swathes of colour. The herbaceous and mixed borders in the walled garden have been planted with a large variety of shrubs and perennials, many of which have a long flowering season. Self-seeders such as pink campion, *Leucanthemum* and *Alchemilla mollis* are allowed grace to stay, when in flower. The two ponds

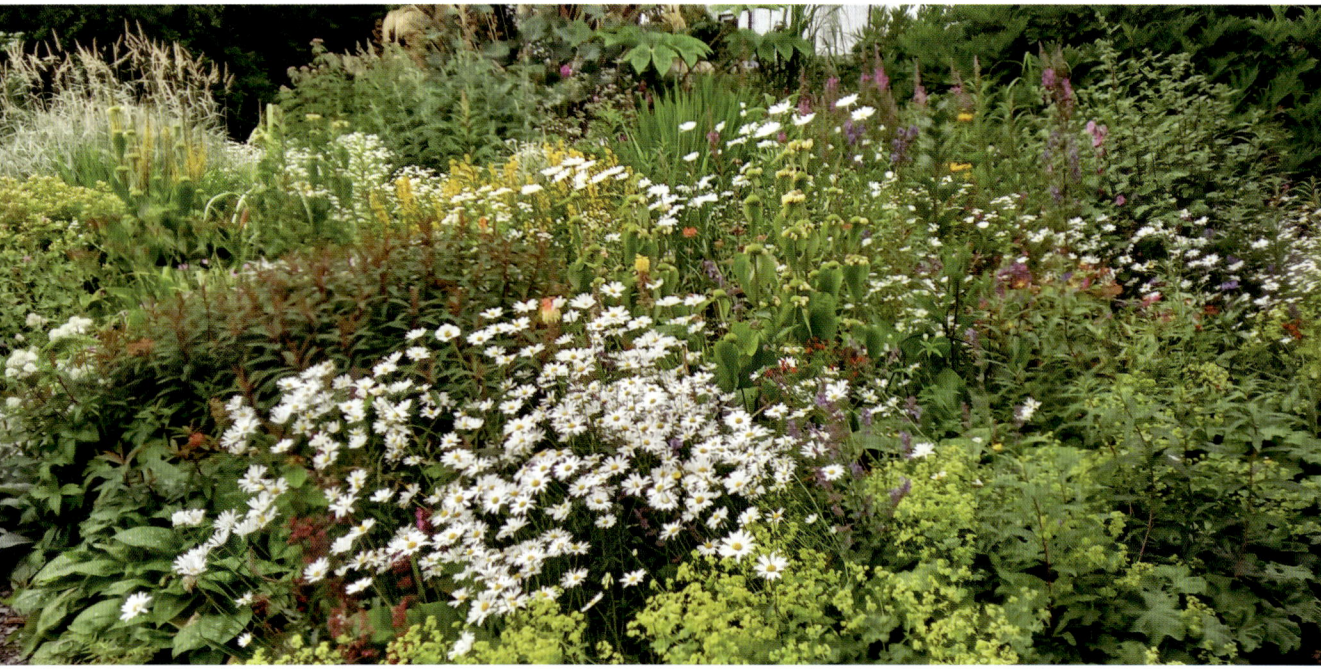

The big perennial bed in the new part of the walled garden at the height of summer.

each have several varieties of water lily and other flowering water plants and the margins are planted with flowering perennials. Planting is dense, and due to the climate, growth is lush, of both wanted flowers and weeds. Families of house martins nest all around the corbels of the tower and are busy across the walled garden, swooping low over the pond to collect insects. This is when the big and blowsy cottage garden flowers really shine – lupins, delphiniums, antirrhinums, shasta daisies and peonies, with a supporting cast of saxifrage 'London Pride' and many different varieties of perennial geranium. The great gardener Margery Fish advised, 'When in doubt, plant a geranium' and I have taken that to heart. They provide flowers for months.

Autumn

The walled garden stays colourful well into October, with roses flowering into November, and late summer perennials such as dahlias, gladioli and Michaelmas daisies still blooming. Some shrubs have been planted with autumn colour in mind, such as euonymous (spindle), cotinus and rowan and several varieties of acer and cotoneaster. One particular flowering cherry

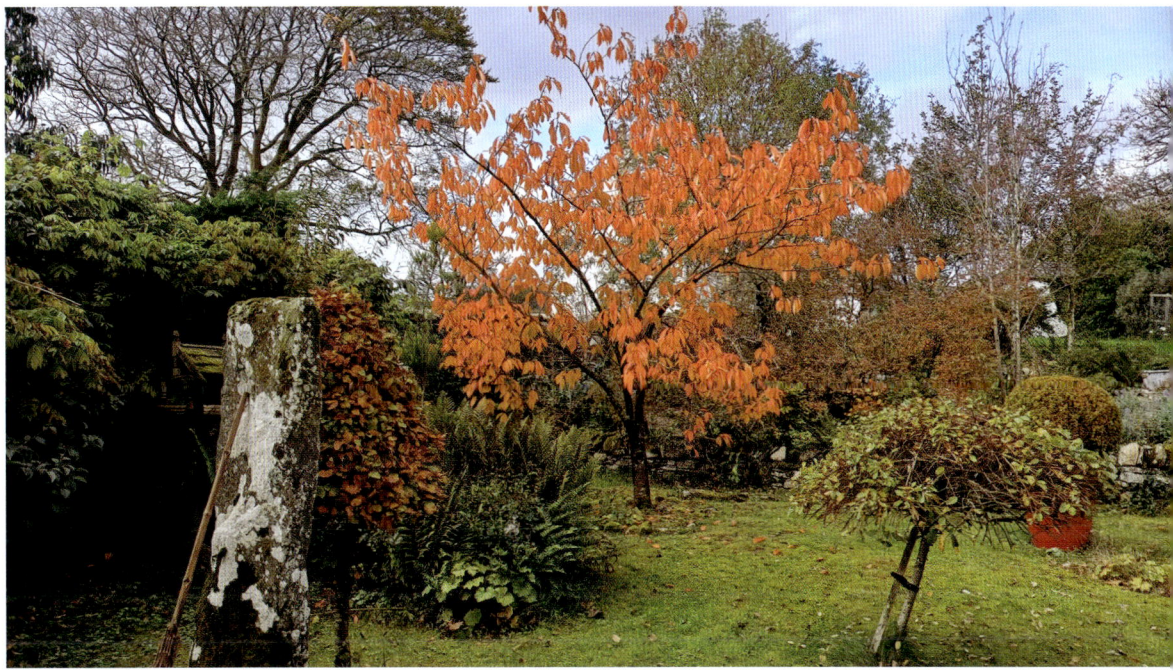

The fluorescent cherry tree in the courtyard in autumn.

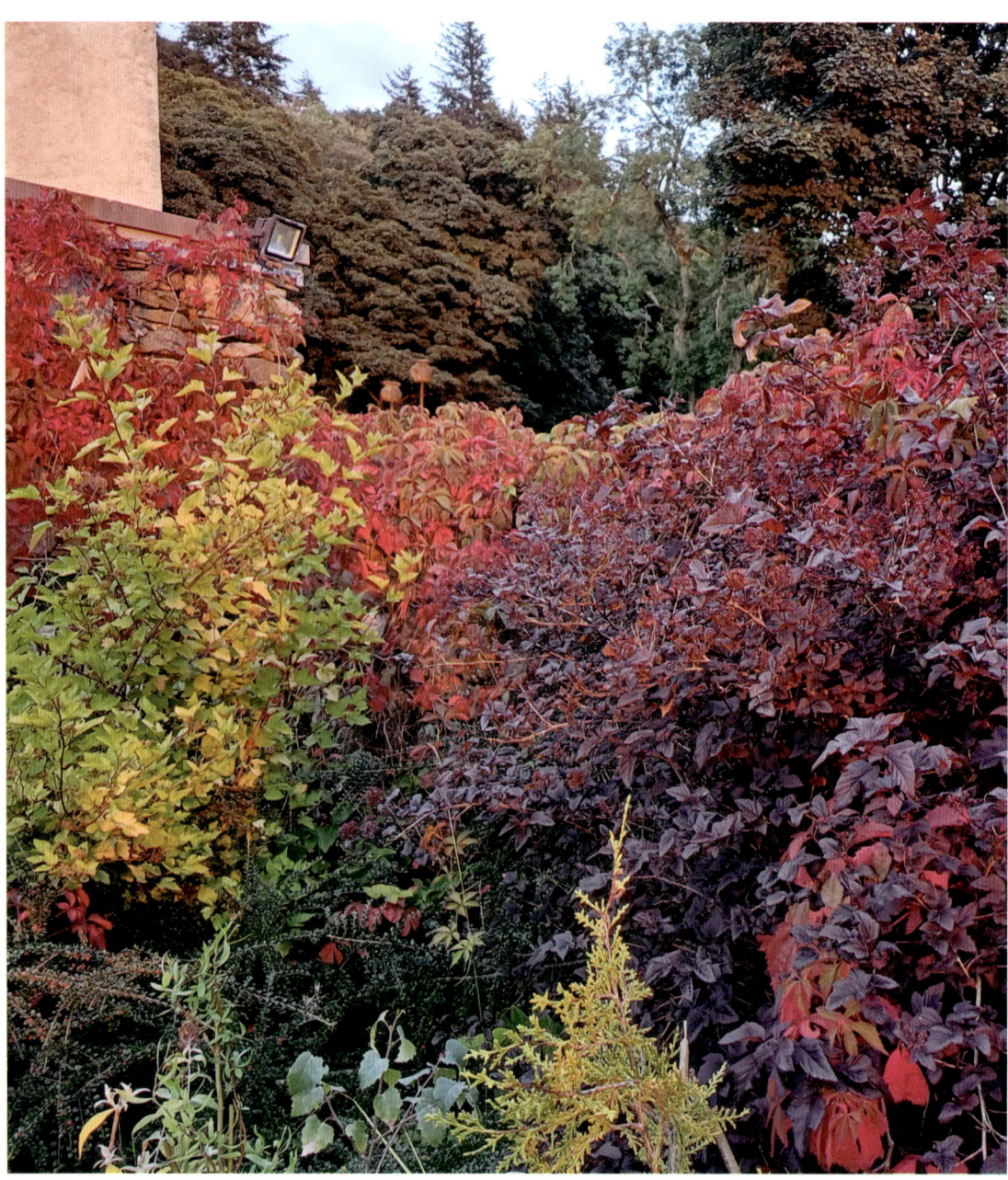

Autumn in the sitooterie, with virginia creeper and *Physocarpus* providing purple and red tones.

in the courtyard has leaves that glow fluorescent orange in October. The traditional autumn colours of red, yellow and orange are complemented by the blues and purples of monkshood, geranium 'Rozanne', catmint and many of our hydrangeas.

THE DEVELOPMENT OF BARHOLM CASTLE GARDEN

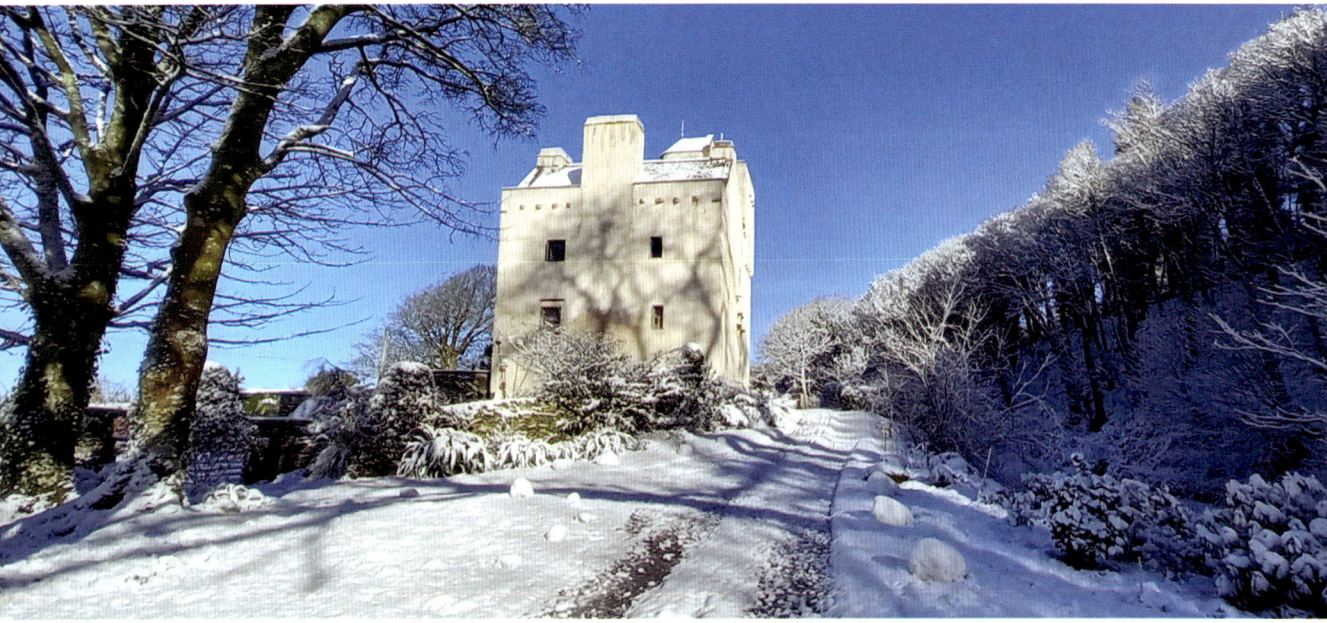

Winter

The garden includes many evergreens, some cut into topiary shapes, to ensure that there is structural interest in the garden all the year round. When we have hoar frost or snow – an increasingly rare event as the climate here becomes warmer and wetter – the topiary shrubs look stunning. There is colour, too, from late January, when the *Chaenomeles* and early camellias start to show deep pink and red and the early-flowering daphnes bring a glorious scent. Seedheads and rough areas are left to encourage birds and wildlife to visit, and we hang nuts and fatballs in the kitchen window recess. In February, hundreds of thousands of snowdrops fill the ravine and signal the coming of spring – and the beginning of the cycle of garden seasons all over again Formerly, the snowdrops were picked by local people, put together in little bunches with an ivy leaf as a contrasting back stop for each, and sent off to the Paddy line steam train for the journey down to Covent Garden, where they were sold for sixpence a bunch. We know this because several times older people have visited and told us that they were snowdrop pickers in the 1940s and 1950s. The Paddy line closed in 1965, a victim of the Beeching railway axe. February also brings the frogs and toads out to mate and lay their spawn in the ponds; we cross our fingers and hope that ice will not kill them off.

A relatively rare but delightful sight – snow at Barholm Castle.

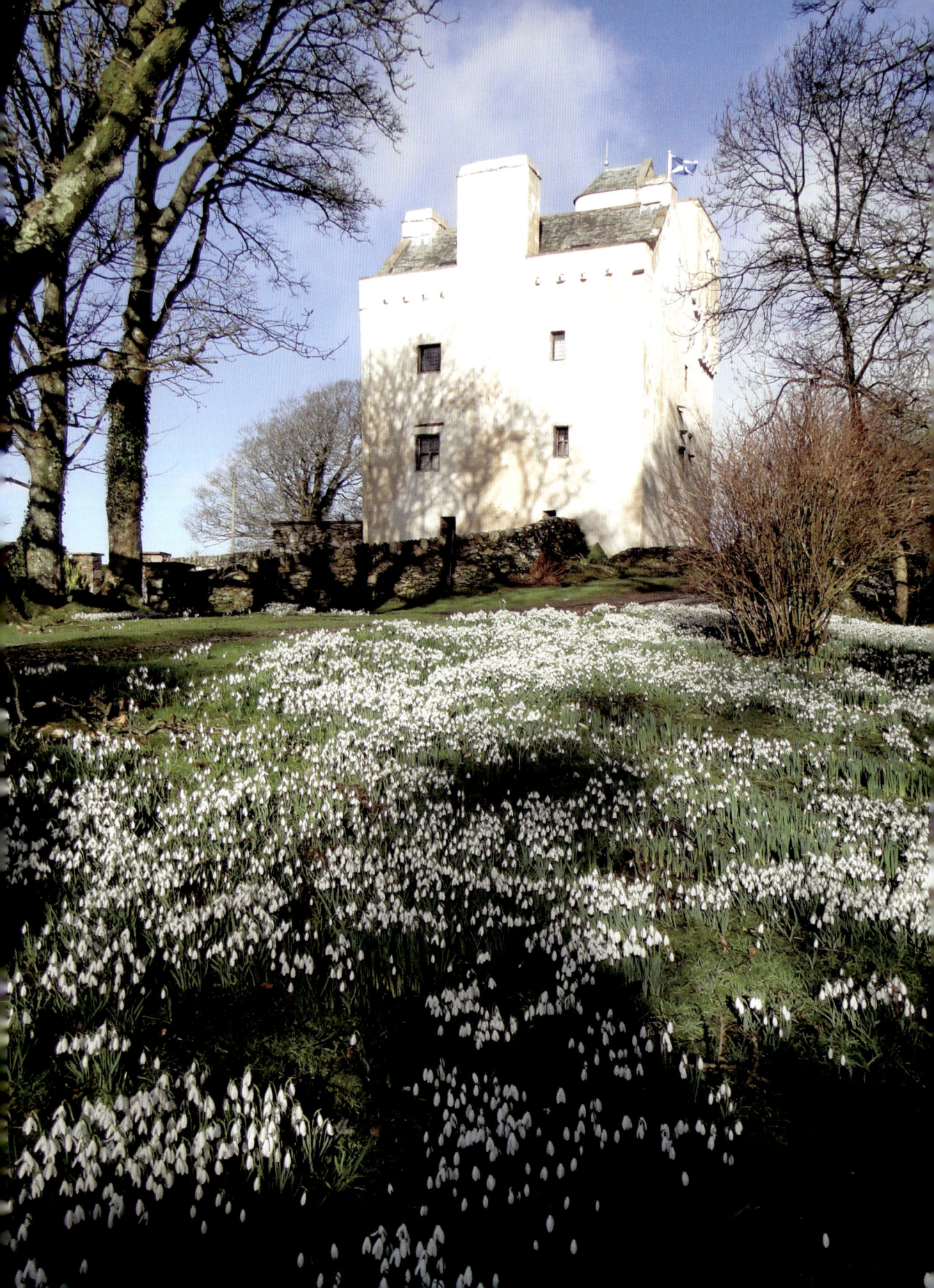

Changing Times in the Garden

Covid restrictions

During the Covid lockdowns of 2020 and 2021 we spent much of our time in the garden, enjoying the peace and tranquillity. We felt very lucky to be confined in such a lovely space with splendid views. Moreover, we had a lot of work to do. We had bought the second half of the walled garden only the year before, and it still needed a massive amount of clearing, weeding and planting. The weather for the first couple of months was glorious, which helped enormously. As restrictions eased, we welcomed visitors to the garden again, at first only in twos and threes, according to the regulations. Some came alone and had been isolating for months. Getting out into a garden and chatting about the plants during their guided tour was a welcome novelty for them. Gradually, groups started to come back; it was liberating to see people out and about again, enjoying themselves.

My garden diary

From the beginning, I kept a note of everything that I planted in Barholm Garden, where I acquired it and where it was planted. As a diary of the garden, it is both uplifting in its record of progress and disheartening to read that so many plants did not survive. All of the experience of loss taught me about what I can and cannot achieve and under what circumstances. Every time that a plant failed to thrive, I had to consider the cause: in the wrong place? Eaten by snails or rabbits? Not enough water? Crowded out by weeds? Too tender to survive winter temperatures or strong winds? In the early days, before we moved to Barholm, many of these potential causes of death were out of my control. Once we moved in 2011, I could monitor delicate plants, keep weeds away from them, provide winter protection if necessary, give adequate water in the first vulnerable months and so on. I was also able to get to know the soil structure – very different in different areas of the garden – and make more judicious choices about where to position plants. I learned a lot.

Keeping a record of where each plant came from has helped to create memories of the people who gave me them and the many nurseries and garden centres I have frequented. I look at the magnificent giant red hot poker, *Kniphofia northiae*, for example, and remember buying it from the

Opposite.
Snowdrops on a sunny day.

late Michael Wickenden at the Cally Gardens, or at the white agapanthus and think of my friend Buffy, who gave me a clump when she was clearing her lovely courtyard garden. The plants in my garden that hold memories and associations make a stroll round the flower beds more meaningful and intimately enjoyable.

Every so often in my diary, I would break out from the mere listing of plants acquired and write some heartfelt lines on the state of the garden. Here are some random entries:

June 2008

[Written after we visited from the Netherlands, after a few months away]

After initial despondency (at the terribly dry conditions and the amount of weed infestation) we found most of what has been planted this year still intact, although it was hard work to locate most stuff under masses of nettles, dock, sticky willie and campion. Even the dahlias, which I had initially thought were all gone, seem to have at least 4 survivors in the long bed . . . The garden was not very colourful (the June drop) – we had just missed a great display of oriental poppies, flag irises and rhododendrons, and the giant alliums and peonies will be out in a week or so, as will the white mallow. The pond was looking great once we uncovered it, and John was delighted that the blanketweed has gone. Two white water lilies were flowering and the third (pink?) has a bud.

June 2010

[An optimistic entry from June 2010, the year before we moved back to live at Barholm]

What a difference from June last year! Everything looks marvellous and healthy – so many different plants, so much growth. Too much, really, and I will have to do some serious cutting back next month, but so much better than finding everything gone. The rabbits have finally gone from the walled garden, through myxomatosis and our rabbit proof fences. The paddock may still be a problem, but we can rabbit proof the border with Downie and that will help. The mound is there now, still to have top soil, and the new shed, which is a great help. The pond looks wonderful and the nine silver birches beside it have all taken well. The planners have agreed to the greenhouse in principle, which is great, so we may have it in place just before we move in. We hope to get a beech hedge in the paddock in October, which will break up the vista nicely.

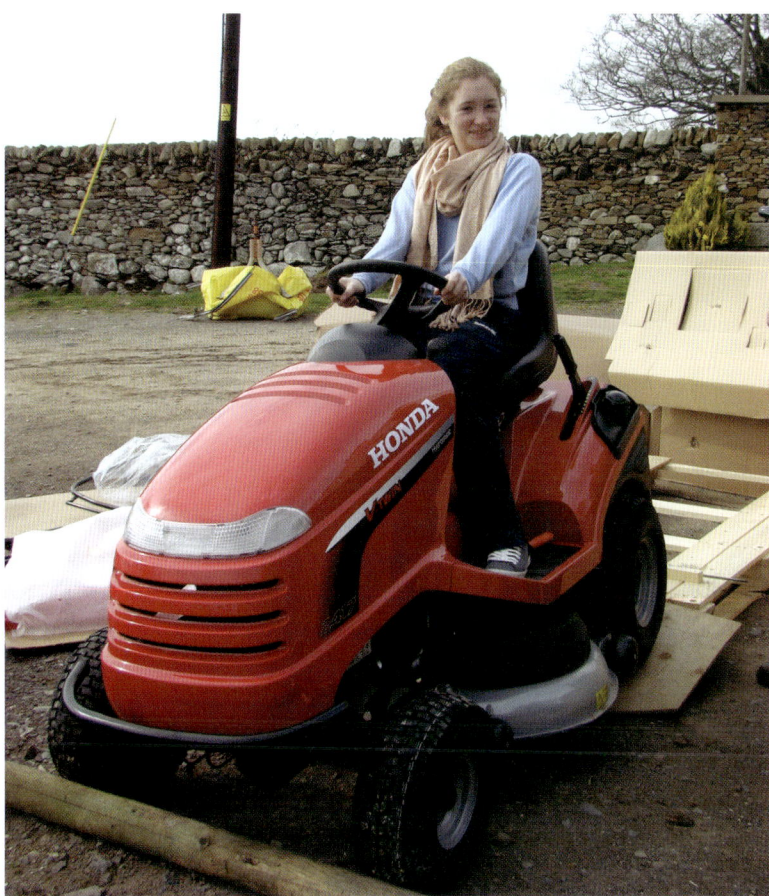

Daughter Rose trying out the newly delivered lawn tractor in 2010. It is still going strong, serviced annually by Ronnie from the Small Engine Clinic in Castle Douglas.

The paddock had its first cut with the new tractor mower this month and already it looks more like a lawn than a field.

April 2012

Everything looking great! Weeding then mulching all the beds with compost for Open Day made a huge difference.

August 2014

Garden looking good – looked at photos from 2012 and am amazed at the difference! The paddock was practically empty and the walled garden didn't have much below the terrace. Tomorrow – plant iris and other by big pond and small pine on east side of castle, and dig over west bed in WG. Put ceanothus in it, for a start. Plant brachyglottis by road near cornus.

April 2018

Hired Ian and his digger and John and Jim helped to cart off the slimy sludge that was the grass heap of the past six years. Then about twenty-five huge rocks were placed next to the heap area, and covered in compost, where I will develop a new rock garden. John P. has started to construct bays for the compost and will finish on Friday, then gradually that area will become much tidier, once the rest of the sludge dries out and we get it seeded. I am excited about planning the big new rock garden – the one beside the castle took relatively little time to become quite mature. I have lots of plans for shifting plants there and also must go and buy plenty more. The situation is ideal and what was a weed-infested dump area of the garden should soon be quite interesting and smart. We are waiting to hear from the solicitor about buying the rest of the walled garden. It will create plenty more work but I hope we get it.

March 2020

Today has felt like the first proper day of lockdown. The men all called off coming to work on the path, so it will probably be months before it is finished. I'm so glad all of the structural stuff is finished in the new bit of the WG, so I can just get on with planting and tending without worrying. I will miss seeing them. However, they have mulched nearly everything in the new bit and what still needs to be done is do-able by ourselves, so we shall just have to crack on. There is plenty of time and it will do us good. The weather was gorgeous today, sunny and warm with no wind. John built a bonfire and we shall start it as soon as the wind is in the right direction. I have got rid of a lot of woody stuff from the new bit and all the plastic pots and junk, so the entrance looks better now. I am so looking forward to the clocks going forward on Saturday night – we shall be able to get out in the garden after tea for months on end!

Caring for the garden

Visitors sometimes seem astonished that we manage such a big garden with very little help, cultivating more than three heavily planted acres. The general help we receive with weeding etc amounts to an average of 2–3 hours one morning per week over the year, plus we have the grass and hedges cut for us by Jim. 'You must be out here working all day every day,' they say. Not so. We lead busy lives and we are both away a lot. However, I do go out with my snippers most days, deadheading and keeping the beds tidy as best I can. I am somewhat ashamed to admit that I do not take particularly good care of my garden tools. I see on television gardening programmes and in magazines that tools are almost fetishised, with stern advice to oil them lovingly and keep them hanging neatly in dry sheds. I do have multiple spades, border forks, trowels and wheelbarrows, but they are kept outdoors, strategically placed around the garden, so that there is always one to hand whenever I need it. The garden is so big that otherwise I could waste a great deal of time and energy walking backwards and forwards searching for what I need. My very favourite tool is a ferociously sharp pair of Jakoti shears (hence my permanently nerve-damaged index finger), used by farmers for shearing sheep and by me for snipping away at anything that needs cutting back. They are like my third hand – I never go into the garden without clutching a pair. They have increased my efficiency in the garden enormously.

I have eight compost bins, ensuring that there is always one nearby for tossing weeds, etc. Nevertheless, there is a lot of walking to be done in such a big garden and 10,000 steps are easily chalked up. There are two wheelbarrows for storing large woody items. These fill up quickly and are wheeled weekly to the massive 20-foot-high bonfire (or 'nonfire') heap which rises discreetly from the ravine to the side of the track. When I have help on a Friday morning, I plan and prioritize the work to be done with a sharp focus on what will make the most visible difference and impact in the garden, such as cutting grass edges, mulching, and clearing weedy patches. A number of areas which are hidden out of the main line of sight are left to look after themselves, more or less. And, of course, I have really great helpers who seem to genuinely enjoy working in the garden and set to with cheerful enthusiasm.

The garden is open every day for Scotland's Gardens scheme, by arrangement. All of the money raised goes to charity, with 60% going to Home-start Wigtownshire, a charity that helps young children and families. We raise hundreds of pounds every year and enjoy showing people around our lovely garden at any time of year.

Garden acknowledgements

We would like to thank those who have helped us with the garden. Although we do the majority of the work ourselves, it has not been possible to do everything, especially the major structural work and keeping on top of things when we were away a lot. Our special thanks go to John Paterson, who directed and carried out work on nearly all of the paths, ponds, pergolas, arbours, gazebos, walls, dykes, sheds etc. and kept the grass mown and hedges cut for twenty-four years. The hard landscaping, which provides such a suitable setting for our planting, was almost all down to John P. Also, thanks to Jim Walker and Brett Self for the amazing paths and steps they made in the ravine and elsewhere. Jim now keeps the grass and hedges expertly cut for us and comes every Friday morning. Over the years I have had help at various times for a couple of hours a week from Abel, Alison, Andrew, Elaine, Eve, Ginny, Katie, Matt, Raymond, Rose, Ruby and Tracy. They have all made a real difference to the garden, supporting me with the chores and, most importantly, keeping my spirits up!

Glossary

anent (Scots) about, concerning

ambry/aumbrey a recess in a wall; in sixteenth-century tower houses these were often used to place lamps

barmkin an enclosure around a tower house for protection

bonnet laird a petty Scottish landowner (who wore bonnets, like the humbler folk, rather than the hats of the gentry; Sir Walter Scott is said to have coined the term in 1816). Bonnet lairds held a position in society below a laird and above a husbandman, or farmer

burn a stream or small river in Scots

Canmore Historic Environment Scotland's website giving details of historic properties

cap house a small room at the top of a spiral staircase

castle for the purposes of this book, a castle is any building that is fortified, or looks fortified, including tower houses

corbel a structure that juts out from a wall to support another structure, such as a beam

cottar a farm labourer or tenant occupying a cottage in return for labour

descheduling removing a building from the HES schedule of ancient monuments which may not be altered in any way

ell an ancient unit of measurement, approximately 37 inches

feuar someone who has the right to use and occupy a piece of land in exchange for paying an annual fee, called feu-duty, to the landowner

garderobe a mediaeval long-drop toilet

harling a rough-cast wall finish made of lime and aggregate

infeft officially in possession of heritable land

laird the owner of a Scottish estate, who ranked below a baron and above a gentleman; it was and is a courtesy title

merk a silver coin used in Scotland in the sixteenth and seventeenth centuries, valued at two-thirds of a Scots pound

motte-and-bailey a type of early castle consisting of a raised earthwork (motte) on which stood a wooden or stone keep, surrounded by a fortified enclosure (bailey)

ogee window a window frame design consisting of two serpentine shaped curves that meet in a point at the top

right of pre-emption right of first refusal, retained by the original seller, when a property or land is put up for sale

roup an auction (Scots)

sasine/seisin an old feudal term for having both possession and title of property.

scheduled ancient monument a nationally important historic building (or site) which is legally protected by being placed on the 'schedule' maintained by HES. There are currently over 8,000 scheduled ancient monuments in Scotland.

wadset mortgage: wadsets were commonly made out in the form of mutual contracts, in which one party sells the land, and the other grants the right of reversion.

wadsetter (Scots law) a creditor to whom a wadset is made

trunk road a road for which the Government, rather than the local authority, has responsibility.

yett a grille of latticed wrought iron bars placed in front of a door or a window, used for defence.

References

Adam, William (2011 [1810]), *Vitruvius Scoticus: Plans, Elevations and Sections of Public Buildings, Noblemen's and Gentlemen's Houses in Scotland*. Dover Publications (written by William Adam in the 1720s, but only published for the first time by his grandson in 1810)

Aitchison, Peter and Cassell, Andrew (2019), *The Lowland Clearances: Scotland's Silent Revolution 1760–1830*. Edinburgh: Birlinn

Blake, Brian (1966), *The Solway Firth*. London: Robert Hale

Crawford, James (2017), 'Cool Scotia: The Great Hall, Stirling Castle', in A. McCall Smith, A. Moffat, J. Crawford, J. Robertson and K. Jamie, *Who Built Scotland?* Edinburgh: Historic Environment Scotland

Cutland, John R. (1986), *The Story of Ferrytown of Cree and Kirkmabreck Parish*. Castle Douglas: Forward Press

Douch, John (1985), *Smuggling – Flogging Joey's Warriors*. Dover: Buckland Press

Fittis, Robert Scott (1903), *Romantic Narratives from Scottish History and Tradition*. Alexander Gardner (available online via the Internet Archive)

Foster, Maurice (2023), *The Hydrangea: A Reappraisal*. Marlborough: Crowood Press

Gordon, J., ed. (1832–45), *The Second Statistical Account of Scotland*. Edinburgh: Blackwood (cited as *Second Statistical Account*)

Harper, Malcolm McLachlan (1876), *Rambles in Galloway: Topographical, Historical, Traditional, and Biographical*. R.G. Mann

Irving, Gordon (1971), *The Solway Smugglers*. Glasgow: R. Dinwiddie and Co.

Langstaff, John Brett (1964), *New Jersey Generations: Macculloch Hall, Morristown*. Godalming: Vantage Press

McCormick, Andrew (1906), *The Tinkler Gypsies*. Gravesend: J. Maxwell and Sons

McCulloch, Andrew (2000), *Galloway: A Land Apart*. Edinburgh: Birlinn

McCulloch, Walter (1964), 'A History of the Galloway Families of McCulloch', unpublished manuscript. Available online at: https://www.scribd.com/document/77675716/A-History-of-the-Gallowat-Families-of-McCulloch

McCullough, Douglas (2025), *A History of Clan McCulloch*, independently published. Available as eBook on Amazon

MacGibbon, David and Ross, Thomas (1887–92), *The Castellated and Domestic Architecture of Scotland*, 5 vols, Edinburgh: Thomas and Archibald Constable

McKean, Charles (2001), *The Scottish Chateau: The Country House of Renaissance Scotland*. Stroud: Sutton Publishing

McKerlie, Peter Handyside (1878), *History of the Lands and their Owners in Galloway*, vol. 4. Paisley: W. Paterson

Macleod, Michael (2024), *Creetown's Industrial Era*. Hold-Fast

McMath, William (1924), *The Gordons of Craichlaw*. Dalbeattie: Thomas Fraser (available online via the Internet Archive)

Maxwell-Irving, Alastair (2014), *The Border Towers of Scotland, Vol. 2 Their Evoution and Architecture*. Published by the author

Morton, Andrew S. (1925–6), 'Barholm Tower', *TDGNHAS*, 3rd Series, 13: 233. (Available online at the DGNHAS website)
Pearce, Michael (2015), 'Approaches to Household Inventories and Household Furnishing, 1500–1650', *Architectural Heritage*, 26: 73–86
Ruskin, John (1873), *The Poetry of Architecture*. London: John Wiley
Russell, J.E. (2000), 'Gatehouse and District, Vol. 1 A Decamillenial History', unpublished manuscript
Sanderson, Margaret (1982), *Scottish Rural Society in the Sixteenth Century*. Edinburgh: John Donald
Scott-Moncrieff, George (1938), *The Stones of Scotland*. London: B.T. Batsford
Sinclair, Sir John, ed. (1791–99), *The Statistical Account of Scotland*, 21 vols. Edinburgh: William Creech (cited as *First Statistical Account*)
Stell, Geoffrey (1996), *Dumfries and Galloway*. Edinburgh: HMSO
Symson, Andrew (2011 [1684]), *A Large Description of Galloway*. Nabu Press
Zeune, Joachim (1992), *The Last Scottish Castles*. Verlag Marie L. Leidorf

Further Reading

Books about Castles, Castle Restorations and Living in Castles

Bailey, Helen (1988), *My Love Affair with Borthwick Castle*. Lewes: Book Guild
Bedford, John, Duke of (1959), *A Silver-plated Spoon*. London: Reprint Society
Bedford, John, Duke of (1971), *How to Run a Stately Home*. London: André Deutsch
Brennan-Inglis, Janet (2014), *Scotland's Castles: Rescued, Rebuilt and Reoccupied*. Cheltenham: History Press
Brennan-Inglis, Janet (2022), *A Passion for Castles: The Story of MacGibbon and Ross and the Castles they Surveyed*. Edinburgh: John Donald
Browne, Nicholas (2007), *Castles and Crocodiles*. Long John Silver Publishing
Burnett, David (1978), *Longleat: The Story of an English Country House*. Glasgow: William Collins
Clow, Robert, ed. (2000), *Restoring Scotland's Castles*. Glasgow: John Smith & Son
Corbett, Judy (2005), *Castles in the Air*. London: Ebury Press
Coventry, Martin (2001), *The Castles of Scotland*, 3rd edition. Edinburgh: Goblinshead
Cruden, Stewart (1981), *The Scottish Castle*, 3rd edition. Glasgow: Harper Collins
Fairbairn, Nicholas (1987), *A Life is Too Short: Autobiography*, Vol. 1 (1987) London: Quartet Books
Ferragamo, Amanda (2005), *Seven Years in Tuscany*. London: Continuum International
Gifford, John (1996), *The Buildings of Scotland: Dumfries and Galloway*. London: Penguin
Guyot, Michel (2007), *J'ai Rêvé D'un Château*. Paris: Éditions J.C. Lattès
Harris, John (1998), *No Voice from the Hall: Early Memories of a Country House Snooper*. London: John Murray
Harris, John (2002), *Echoing Voices: More memories of a Country House Snooper*. London: John Murray
Laing, Gerald (1974), *Kinkell: The Reconstruction of a Scottish Castle*. Dingwall: Ardullie House
Lees-Milne, James (2006), *Diaries, 1942–1954*, abridged and Introduced by Michael Bloch. London: John Murray
Lees-Milne, James (2006), *A Mingled Measure: Diaries, 1953–1972*. Norwich: Michael Russell
Lindsay, Maurice (1986), *The Castles of Scotland*. Edinburgh: Constable
Maclean-Bristol, Nicholas (2007), *From Clan to Regiment: Six Hundred Years in the Hebrides 1400–2000*. Barnsley: Pen and Sword (includes a first-person account of the restoration of Breachacha Castle)
Macpherson, Sandra (2004), *A Strange and Wild Place*. Edinburgh: Birlinn
Queen Marie of Roumania (1925), *The Country That I Love – An Exile's Memories* (Chapter 15: Brana the Beloved). London: Duckworth
Miller, Christian (1989), *A Childhood in Scotland*. Edinburgh: Canongate Classics
Montagu of Beaulieu (1967), *The Gilt and the Gingerbread: or How to Live in a Stately Home and Make Money*. London: Sphere Books

Murdoch, Ken L.S. (2010), *Methven Castle: The Restoration of a Seventeenth Century Building*. Perth: Ken L.S. Murdoch
Parris, Matthew (2005), *A Castle in Spain*. London: Viking
Rathbone, Belinda (2005), *The Guynd: A Scottish Journal.* New York: Quantum Lane Press
Renucci, Martin (2011), *Guedelon: A Castle in the Making. Rennes:* Ouest France
Rolland, L.A.L. (1977), *Rossend Castle*. Published online at: http://www.brand-dd.com/burntisland/rossend2.html
Ross, Susan (1973), *The Castles of Scotland*. London: Letts Guides
Roy, James Charles (2003), *The Fields Of Athenry*. Oxford: Westview Press
Saddlemyer, Ann (2002), *Becoming George: The Life of Mrs W. B. Yeats*. New York: Oxford University Press
Salter, Mike (1985), *Discovering Scottish Castles*. London: Shire Books
Strachan, Sabina Ross (2008), 'The Laird's Houses of Scotland: From the Reformation to the Industrial Revolution, 1560–1770', PhD thesis, University of Edinburgh
Tranter, Nigel (9165), *The Fortified House in Scotland, Vol. 3: South-West Scotland*. Edinburgh: Mercat Press

Miscellaneous Books

Chatto, Beth and Lloyd, Christopher (2021), *Dear Friend and Gardener*. London: Frances Lincoln
Dick, C.H. (1927), *Highways and Byways in Galloway and Carrick*. Basingstoke: Macmillan
Goodman, Ruth (2016), *How to be a Tudor*. London: Penguin Books
Jekyll, Gertrude (2019 [1908]), *Colour in the Flower Garden*. Home Farm Books
Robertson, Forbes W. (2000), *Early Scottish Gardeners and Their Plants 1650-1750*. East Linton: Tuckwell Press
Sayer, Dorothy L. (1949), *Five Red Herrings*. London: Victor Gollancz
Scott, Sir Walter (2003 [1815]), *Guy Mannering*. London: Penguin Classics (first published, anonymously, in 1815)
Smit, Tim (2010), *The Lost Gardens of Heligan*. London: Victor Gollancz
Strong, Roy (2003), *The Laskett: The Story of a Garden*. London: Bantam Press

Cato, Susan, Betty, George and Emma, enslaved people, Macculloch Hall, 58
Cawdor Castle, 35
Celtic castles, 102
Charles I, king, 40
Charters, James of Dalry, 20
chimney at Barholm, 28, 83
Cholera, 60
Cibber, Colley, 75
Claughreid estate, 41, 42
Compost, 140, 142
Compstone (Cumstoun) Castle, 17
Conchieton, 25
Conventicles, 40
Copland, John (artist), 105
costs/funding of restoration work (Barholm), 79–82
Count de Lauriston, 58
Covenanters, 40,
Covenanting tradition, 69
Covid lockdown, 34, 105, 120, 179, 182
Cox, Ken, 70
Craigcaffie Tower, 11, 14, 18
Craigievar Castle, Aberdeenshire, 4
Crawford, James, 101
Cree estuary, 47
Cree Landscapes (garden company), 122–3
Creetown, 3, 20, 42, 43, 45, 47, 48, 49
Creetown's Industrial Era, 36
Crichton-Stuart, John, 3rd Marquess of Bute, 18
Crichton-Stuart, John, 4th Marquess of Bute, 18
Crockett, S.R., 18
Crothers, Michael, 70
Crothers, Sarah, 70
Cruden, Stewart (historian), 10
Cruggleton Castle 13, 17
Cullintnick, William, 38
Culzean Castle, 47, 100
Cumming and Co., 79
Cumstoun Castle, 17
Cutlar, Elizabeth of Argrennan, 58

Dalbeattie, 17
Davies, Heather (artist), 105, 106-7
digital camera, 77
Dirk Hatterick's Cave, Wigtown Bay, 19
door of Barholm Castle, 30, *31*
Doors Open weekend, 108, 139
Double Gallant, The (play), 75
Douch, John, 60

Dowies (Old Place of Monreith), 11, 14, 18
Drum Castle, Aberdeenshire (door), 30, *31*
Drumcoltran Tower, 178
Drumlanrig Castle, 47, 97
Drummond, Peter (architect), 82, 86
Duart Castle, Mull, 100
Duke of Atholl: see Atholl, 2nd Duke of
Dumfries, 60, 82
Dumfries District Board of Control, 65
Dumfries House, 170
Dumfries and Galloway Natural History and Antiquarian Society (DGNHAS), 69
Dundeugh Castle, 18
Dundrennan Abbey, 32, 115
Dunskey Castle, 14, 17
Dutch elm disease, 121, 146
dykes, 21, 27, 32, 121, 124, 128, 139, 152, 158

Earlstoun Castle, 17
Edinburgh, 48, 58, 60
Edinburgh Castle, 86
Edingham Castle, 17
'Ellangowan' (castle in Walter Scott's *Guy Mannering*), 67
Elshieshields Tower, 7
Extreme Homes (television programme), 80

Faed, James (artist), 105, 106
Faed, John (artist), 105
Ferguson, Alexander of Kilkerran, 54
Ferry Toun (Ferrytown) of Cree, 3, 41
fireplaces in Barholm Castle, 84
First Statistical Account, 45, 47, 52
First World War, 65, 118
Fish, Margery (gardener), 175
Fittis, Robert Scott, 54
Five Red Herrings (Dorothy L. Sayers),1 03
fruit and vegetables (Barholm), 147

Galdenoch Castle, 14
gallery, 7, 8
gallery at Barholm, 39, 95-6
Galloway, Earl of, 25
Galloway castles, 10–18
garden benches (Barholm), 153-5
garden gates (Barholm), 102, *103*, 125, 153
Garden of Cosmic Speculation, Dumfries and Galloway, 155
garden paths (Barholm), 151-2

Index

Page references in italic are to illustrations.

A75 trunk road, 36, 86, 109, 120
Abbot's Tower, 11, 18
Adam, Robert, 36, 43, 49
Adam, William, 43
Ainslie, John, Map of the County of Wigton 1782, 66, 67
Aitchison, Peter, 21
Anwoth, 19, 54
Archibald the Grim, 17
Atholl, 2nd Duke of, 19
Auchenskeoch Castle, 17
Auchness Castle, 11, 15
aumbrey, 84, 85

Balbegno, 26
Baldoon Castle, 14
Balhassie (Balhazy), 42, 45
Balmangan Tower, 17
Balzieland Castle, 15
Barclosh Castle, 17
Bardristane, 43
Barholm Byre, 70
Barholm Farm(house), 64, 69, 70, 126
Barholm House (1960s building near Barholm Farm), 70, 71, 76, 79, 130, 132, 159
Barholm House, Creetown, 19, 43-5, 46, 47-9, 50, 51, 52, 57, 60, 62, 63, 64, 65, 66
Barholm Mill, 70, 71
Barmkin, 3, 4, 18
Barrie, Janet, 70
Barrie, Will, 70
Barscobe Castle, 11, 17, 18, 26
Battle of Bothwell Bridge, 41
Battle of Langside, 17
Battle of Rullion Green, 40
BBC, 79
Bedford, Duchess of (Mary du Caurroy Tribe), 57, 63
Bedford, 12th Duke of (Hastings Russell), 56, 63-4
beds in Barholm Castle, 94, 95, 96, 97
Begg, Ian (architect), 75

Bek, John in Kirkbryde, 39
Benmore Garden, Argyll, 70
Berenice (Capt. Grant's steam frigate), 62
biodiversity, 136-8
Blake, Brian, 157
Bloom, Alan (horticulturalist), 164
bluebell wood at Barholm, *120*
Bonnie Bess (Elizabeth McCulloch of Barholm), 56
Booth, Adam (artist blacksmith), 102, 125
Brennan, John, 82, *86*, 129, *130*, 142, 167, 182
Brennan, Rose, 82, 84, 108, *181*
Bressingham Gardens, Norfolk, 164
Brexit, 118, 169
Briggs, Andrew (artist), 102, *103*
Brodie Castle, 100
Brown (Broun), John of Carsluith, 28, 39
building warrant, 78, 81
Buittle Castle, 11, 12, *16*, 17, 18, 47, 115
Burgh of Barony, 49
Bussabiel (Bush o' Bield), 54, 56

Cairn Holy, 69, 157, 158
Cairnsmore House, 63
Cally Castle, 11, 16
Cally Gardens, 135, 149, 180
Cambret, 42
Canmore, 14, 43, 102
Cardoness/Cardiness, Lady (Marion Peebles, widow of Gordon of Cardoness), 54-5
Cardoness Castle, 11, 15
Carsluith Castle, 12, 26, 28
Carson, Reverend, 19
Cassell, Andrew, 21
Cassencarie Castle (Castle Cary), 12
Cassencary House, 19
Castle Clanyard, 11, 14
Castle Douglas, 17
Castle Huntly, 41
Castle Kennedy, 14
Castle of Park, Glenluce, 11, 14, 18, 47
Castle of St John, 14
Castle Stewart, 12

Union of the Parliaments 1707, 19

VAT on building costs, 79
Victorian farms, 2
Vitruvius Scoticus, 43, 45
Vivat Trust, 18

Walker, Jim, 113, *120*, 144, *150*, 169, 184
walled garden (Barholm), 35, 102, *106*, 112–3, *116*, 118, *119*, 121, *122*, 124–7 143, 148, 154, 155, 169, *172*, *173*, 174–5
water (Barholm Castle and garden), 35, 142–3
weeds (Barholm), 135–6
Weekes, Frederick Wickham, 51, 64, *65*, *66*, *67*
Whitford, Patrick, 71
Whithorn, 3, 153

Whithorn Priory, 115
Wickenden, Michael, 180
Wickham Place (Barholm House stables), *51*
Wigtown, 3
Wigtown Bay, 19, 47, 48, *112*, 113, 124, 128
windows, 8, *10*
windows of Barholm Castle, 31–2, 93
Woburn Abbey, Bedfordshire, 63
women in the sixteenth century, 4
women's suffrage, 61
Wood, Andrew, 26

'Yellow Book' (Scotland's Gardens), 143, 183

Zeune, Joachim (castle historian), 28

INDEX

Newton Stewart, 12, 13
Nowell, Max, 102, 105

Old Place of Mochrum, 12, 18
Old Place of Monreith, 11, 14, 18
Orchardton Tower, 17
Ordnance Survey, 102
Orroland House, 32

Paddock (Barholm), *128–9*
painted ceiling in Barholm Castle, *92, 93*
Palnure, 12
Park House: *see* Castle of Park
Partick Castle, 33–4
Paterson, John (John P.), 118, 123, 140, 184
Pearce, Michael, 6
Peebles, Marion (Lady Cardoness/Cardiness), 54–5
Pentland Rising, 40
pests and diseases, 140
plans and sections of Barholm Castle, *33, 77*
Pont, Timothy, 25
Porter, Walter of Blaiket, 25
Portland cement, 100
Portpatrick, 14, 152
Privy Council, 40, 55
propagating, 145–7
pruning and topiary (Barholm), 160–1
Pybill estate, 41

Rabbits (Barholm), 138–9
Rambles in Galloway, McLachlan Harper, 2, 66
ravine (Barholm), 129–31
RCAHMS, 84
RCAHMS Inventory 1914, 24
Reformation, 10
Regent Moray, 17
Rhododendrons (Barholm), 70, 121, 124, 130, 149, 150, 174
right of pre-emption, 76
Rizio, David, 52
rockeries (Barholm), *141, 166–7, 166, 167*
Romantic Narratives from Scottish History and Tradition, R.S. Fittis, 54
Ross, Pat (artist), 102
Royal Botanic Garden, Edinburgh, 70
Rusco Tower, 11, 16, 18, 47, 87
Ruskin, John, 171
Russell, J.E., 24, 37
Rutherford, Samuel, 54

saltire flag (Barholm), 113
Sanderson, Margaret, 2
Sanquhar Castle, 18
Saunderson, Louisa Edwina, 58
Sayers, Dorothy L., 105
scheduled monument, 69, 75, 76, 78
Scotland's Gardens ('Yellow Book'), 143, 183
Scott, Captain Thomas, 56
Scott, Sir Walter, 56, 67, 105
Scott-Moncrieff, George, 42
Scotts Castle Holidays, 102
sculpture (Barholm), 30, *156, 157*
Second Statistical Account, 45, 63, 157
Second World War, 70
self-seeding plants (Barholm), 148–9
Shirreff, Elizabeth Davidson, 56, 57
Shirreff, James, 56, 57
Smailholm Tower, 4
Smit, Tim (gardener), vii
smuggling, 18–19, 47
Snape, Martin, 105
Sorbie Tower, 12, 14
spiral staircase (Barholm), 25, *87*
Spong, Ambrose Henry, 51, 66
Spong, Gill, 66
St John's Town of Dalry, 17
stair tower (Barholm), 25, 27, 30–1, 32, 35, 39, 87–8, 96
'standing stones', 157, *158, 159*
Stell, Geoffrey, 75
Stewart, Agnes, 60
Stirling Castle, *101*
storm damage, 149–51
    Storm Arwen, 149
    Storm Éowyn, 151
    Storm Kathleen, 150
Stranraer, 14
Strathmore, Earl of, 41
Strong, Roy (gardener), vii, 113
sunken garden, 120, 129, *132–3*, 152, 167
Sweetheart Abbey, 18, 115
Symson, Revd Andrew, 14, 25

Temperance Society, 61
Threave Castle, 12, 17
Tongland Abbey, 115
Tudor Monastery Farm (television series), 2
Twynholm, 17

195

Lloyd, Christopher (gardening writer), vii
Lochnaw Castle, 11, 12, 14, 18, 47
Logan Botanic Garden, 15
Lord Lyon Office, 51
Lowland Clearances, 20–1
Lutyens, Edward, 113

Macculloch Hall, Morristown, New Jersey, USA, 58
MacGibbon, D. and Ross, T., 10, 11, 14, 17, 26, 32
Machermore Castle, 11, 12, 18
Mackintosh, Charles Rennie, 101
Maclellan, Donald of Gelston, 25
Maclellan's Castle, 17
MacLeod, Michael, *Creetown's Industrial Era*, 36
Manxman's Prayer, 37
Marquess of Bute (3rd and 4th), 18
Mary, Queen of Scots, 52
master bedroom in Barholm Castle, 93, 94
Maxwell family 11
Maxwell, Achilles John, 28
Maxwell, Laird, 39
Maxwell, Sir John, 4th Lord Herries of Terregles, 17
Maxwell-Irving, Alistair, 9, 25, 26, 28
Maybole Castle, Ayrshire, 32
McClelland, Thomas of Gelston, 25, 38
McCulloch, Alexander of Ardwall, 37
McCulloch, Andrew, author of *Galloway: A Land Apart*, 3, 41
McCulloch, Baronet of Myretoun, 54
McCulloch, Captain William ('Flogging Joey'), 59
McCulloch, David, 25, 28, 37, 38, 39
McCulloch, Elizabeth (mother of Freddie Weekes), 64, 65
McCulloch, Elizabeth of Barholm (Bonnie Bess), 56
McCulloch, Francis Law, 58
McCulloch, George, 48
McCulloch, George Perrott, 57
McCulloch, Godfrey, 49, 54, 56
McCulloch, Harry, Younger of Barholm, 54
McCulloch, Henry, 41
McCulloch, Isabella, 51, 57, 61, 62, 64
McCulloch, James II, 28
McCulloch, James Murray, 60
McCulloch, James of Bardristane, 37, 39
McCulloch, James of Cardoness, 25, 38, 39
McCulloch, John I of Barholm, 25, 37, 39
McCulloch, John II (Major) of Barholm, 40, 41
McCulloch, John III of Barholm, 41, 42
McCulloch, John IV of Barholm, 41, 42, 43, 47, 48
McCulloch, John V of Barholm, 43, 45, 48, 51, 60
McCulloch, John VI of Barholm/Balhasie, 48, 49, 51, 57, 60, 61, 62
McCulloch, John of Conquhieton, 37
McCulloch, John of Myretoun, 54
McCulloch, Mary Louisa, 58
McCulloch, Thomas, 28, 39
McCulloch, Thomas of Cardoness, 25, 37, 39
McCulloch, Walter, 38, 40, 54
McCulloch coat of arms, *49*
McCulloch Grant, Agnes, 51, 57, 62, 63, 64, 65, 67
McCulloch Grant, James, 51, 62, 64
McCulloch Grant, Jane, 51, 57, 62, *64*, 65
McCulloch Grant, John, 51, 64
McCullochs of Muil, 51
McCullochs of Myretoun and Cardoness, 51
McDowell, Isobel, 41, 42
McKean, Andy (artist), 102, 106
McKean, Charles (historian), 11, 24
McKerlie, Peter Handyside (historian), 61, 62, 66, 67
McKie, Alexander of Broach, 39
McKie, Mary (wife of Thomas McCulloch), 39
McLachlan Harper, Malcolm, 2, 66
Menzies, Margaret (wife of David McCulloch), 39
Merredew, Jennifer, 92
Ministry of Works, 100
Monreith House, 37
Morrell, Tony (Blacksmith), 88
Morris County, New Jersey, USA, 58
Morton, Andrew S., 25, 38, 54, 55, 69
mound, the (Barholm), 155, *156*
Muir, Marion, 38
Muir, Revd John, 63
Muirfad Castle, 12
mulching, 182
Murray, James, 2nd Duke of Atholl, 19
Murray of Cally, 45
Musky the farm cat, *151*, 153
Myrton Castle, 11, 12

Napolean Bonaparte, 58, 59
(National) Covenant, 40
National Trust for Scotland, 17, 101
Netherlands, 28, 112, 118, 149, 155
New Abbey, 18
New Galloway, 17
New York, 58

INDEX

garden ponds (Barholm), 161, *162*, 163
garden sculpture (Barholm), 156–7
garden thugs (Barholm), 134–5
Gardeners' Question Time (radio programme), 169
garderobes, 7, 8, 9, 97
Garlies Castle, 13
Gatehouse of Fleet, 4, 16, 42, 87
Ginger, the Brennans' cat, *157*
Glamis Castle, Angus, 41
Glenluce Abbey, 115
Glenwhan Garden, Stranraer, 113
Glorious Revolution 1688, 40, 41
Gordon, Alexander of Cardiness, 54
Gordon, Alexander of Kirkcowan, 42
Gordon, Jean of Culvennan, 41, 42
Gordon, John of Culvennan, 41
Gordon, Nathaniel of Kirkcowan, 42
Gordon, William of Cardiness, 54, 55, 56
Gordoun, Margaret, 39
graffiti (Barholm), 68
Grant, Captain George, 51, 58, 61, 62, 64
grass and lawns (Barholm), 143–4
Great Dixter Gardens, East Sussex, vii
Great Hall in Barholm Castle, 8, 9, 10, 28, 36, 39, 76, 84–6, 90, *91*, *92*, 93, 102, 106
greenhouse (Barholm), 144–5
Greyfriars Churchyard, Edinburgh, 40
Guy Mannering (Walter Scott), 56, 67, 105

Hannay, Colonel Frederick Rainsford, 69, 70, 76
Hannay, David, 24, 70
Hannay, Major Ramsay Rainsford, 69, 70
Hannay, Sir Samuel, 36, 45, 48
Hannay, William, 42, 43
harling, 99, 100
hedges and trees (Barholm), 159–60
Heidelberg Castle, Brunnenhalle 32
Heligan Gardens, Cornwall, vii
Herries, Agnes, Baroness of Terregles, 17
Hewison, Revd Dr J. King, 69
Hewitt, Claire (artist and mapmaker), 102, 105
Highland Clearances, 20
Hill House, Helensburgh, 101
Hills Tower, 11, 18
Historic Environment Scotland, 85
    Inventory of Designed Landscapes, 113
Historic Scotland, 76, 78, 81, 82, 84, 88, 100
    grant, 76, 78
Hogg family, 42

Hogg, Ann, née Younger, 70, 71
Hogg, Billy, 70, 71
Homes by the Sea (television programme), 80
Home-Start, Wigtownshire (charity), 183
house martins, 99, *137*, 175
Hughan, Andrew (footman), 52
Hyslop, Rab (blacksmith), 153

Ireland, 48
Irving, Gordon, 19
island beds (Barholm), 164, *165*, *171*
Isle of Man, 19, 43, 47, 60, 96
Isle of Whithorn Castle, 11, 14, 102

James V, king of Scotland 3
Jekyll, Gertrude, 113
Johnson, MacCulloch and Law, 58, 59

Kenmure Castle, 17, 30, 47
King George V playing fields, Creetown, 65
Kirkclaugh House, 19
Kirkconnell Tower, 11, 18
Kirkcudbright, 17, 40
Kirkdale Archaeology, 26, 28, 82
Kirkdale Bridge, 36
Kirkdale Church, 62
Kirkdale Estate Trust, 71
Kirkdale House, 36, 45, *46*, 66
Kirkdale Sawmill, 70
Kirkpatrick, Elizabeth (mother of Thomas McCulloch), 28, 39
Kisimul Castle, Isle of Barra, 100
kitchen in Barholm Castle, *88*, 89
Knott, Tessa (gardener), 113
Knox, John, 39, 52, 53, 67
    John Knox's room (Barholm), 96–7

Lafone, Mary Ellison, 60
Lanark, 40
Landmark Trust, 14
Langstaff, John Brett, 58
Laskett, The (garden), Herefordshire, vii, 113
Law, Francis, 58
Lennox Plunton 4, *15*, 16–17
Levellers, 21
Limerich (building conservation company), 99
limewash, 12, 99
Lindsay, Ian (architect), 47, 76, 77
Livingstone, Andrew, of the Airds, 56